BrainSnack®

Enjoy thinking!

imagine!
Publishing

Frank Coussement & Peter De Schepper

Brain Twisters, Mind Benders, and Puzzle Conundrums

With
BrainSnack®
puzzles

imagine!
Publishing

Library of Congress Cataloging-in-Publication Data Is Available

10 9 8 7 6 5 4 3 2 1

An Imagine Book
Published by Charlesbridge
85 Main Street
Watertown, MA 02472
617-926-0329
www.charlesbridge.com

Manufactured in China, January 2011

ISBN 13: 978-1-936140-29-9

For information about custom editions, special sales, premium
and corporate purchases, please contact Charlesbridge
Publishing, Inc. at specialsales@charlesbridge.com

BrainSnack® and Binairo™ are registered trademarks of Peterfrank t.v. Belgium.
A Binary puzzle is a PeterFrank puzzle also known as Binairo™.

PeterFrank t.v.,
Postbus 11
B-9830 Sint-Martens-Latem, Belgium
www.peterfrank.be

Contents

Solution strategies

To solve a BrainSnack® puzzle you will have to apply one or more of the following solution strategies. The most important solution strategy is mentioned for many BrainSnack® puzzles.

 Look for the elements in the BrainSnack® puzzle that differ from the rest.

 Look for similarities and associations between the different elements.

 Colors are important.

 Pay attention to the three-dimensional design.

 Calculate with the digits and numbers that you see and/or discover.

 Elements are shifted or follow a logical direction.

 Look for the logical order or a repetitive series.

 Try to find the logic.

Degree of difficulty

From easy to difficult

Brain Sport at Its Best

Compared to our weight on the scales, the brain is a mere trifle; a discountable gray mass that only weighs about 2 percent, but it's a whopper! The brains in your head are the source of knowledge, logical thought, imagination, creativity, emotion, and memory. In short, everything that's important in life.

It's important to realize that the brain is something dynamic. It isn't static, it can be trained. Your gray matter continues to make new cells at every age. This enables you to continue developing at all levels. In other words, your capacities may be partially innate but you can perfect them via exercises. You can learn to use your capacities and excel.

To solve all the fun BrainSnack® puzzles—logical brainteasers, memory games, word problems, concentration exercises, raster puzzles, and more—in this book, your hundred billion brain cells will have to lend a helping hand.

Tip!
Solve these puzzles in a well ventilated room because your brain uses about 20 percent of all the oxygen that is pumped through your body.

Brain-teasers

You can see
me yet I am
weightless.
Put me in a
bag and I make
it lighter.
What am I?

Solution on page 288

Left and right in balance

The human brain still hasn't revealed all its secrets. We do know that your left and right cerebral hemispheres work together via a bundle of nerve cells that link the hemispheres. Logical and pragmatic word problems challenge your left hemisphere a bit more. Creative riddles that force you to think out of the box stimulate the right hemisphere a bit more. Although everyone has two hemispheres, most of us have developed a slight preference for one hemisphere or the other. Everyone enjoys BrainSnack® puzzles since they stimulate both hemispheres.

The left hemisphere deals with more	*The right hemisphere deals with more*
Logic	Fantasy
Analysis	Synthesis
Rationality	Intuition
Order	Color
Linearity	Three-dimensionality
Detail	Associations
Numbers	Images
Words	Rhythm
Seriousness	Humor

Brainteasers

At the bottom you see the strip of images that is glued
to each of the five wheels of this gambling machine.
Which symbol (1–6) should replace the question mark?

Brainteasers

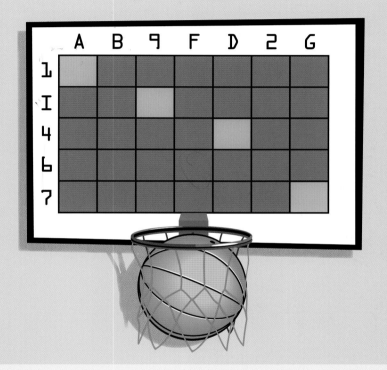

A point is scored when the ball lands on an orange square. Give the coordinates of the last square that should be orange. Answer like this: 4B.

Brainteasers

Which letter should replace the question mark?

Brainteasers

4

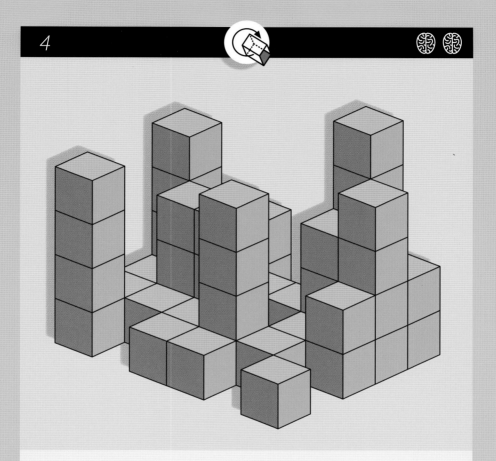

What is the minimum number of blocks that you need to re-create this construction from scratch?

Brainteasers

5

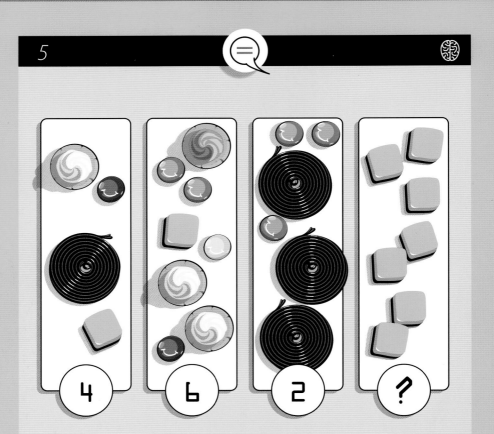

Which number should replace the question mark?

Brainteasers

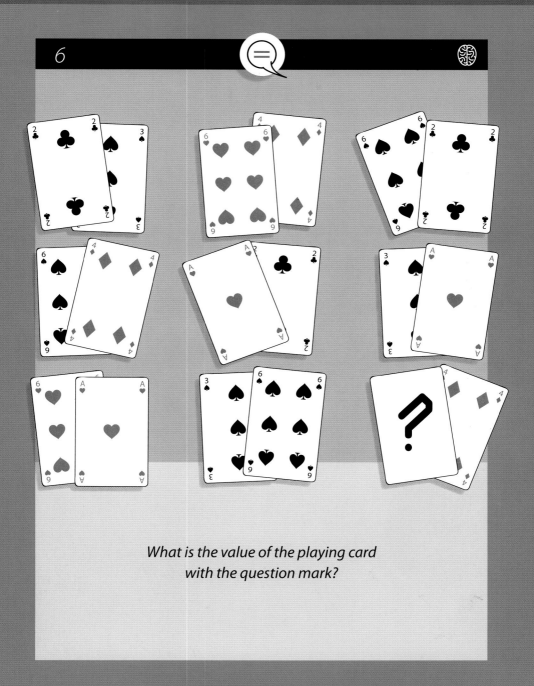

*What is the value of the playing card
with the question mark?*

Brainteasers

*How many dots should replace
the question mark on the domino?*

Brainteasers

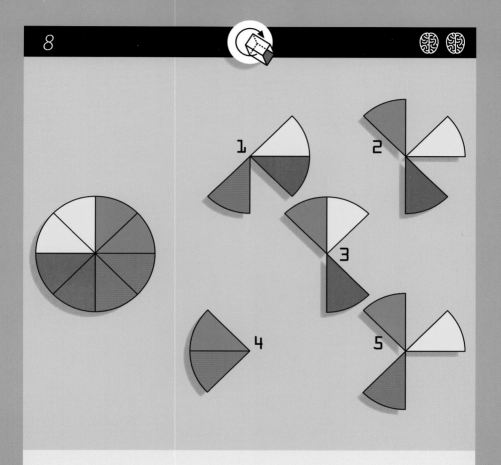

Which three elements (1–5) are needed
to create the entire pie chart?
Watch out, all the elements are shown as mirror images.
Answer with the numbers in increasing order, e.g. 124.

Brainteasers

9

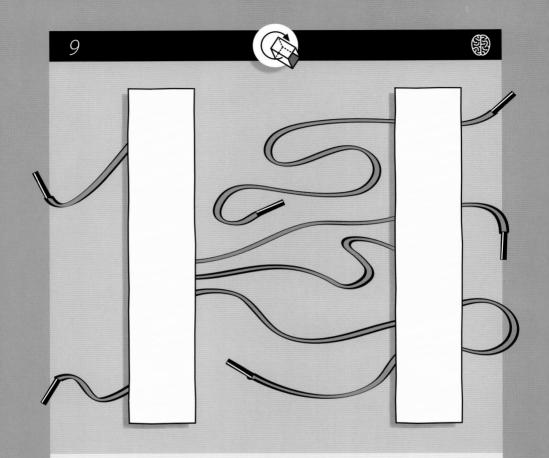

How many shoelaces are shown here?

Brainteasers

Which train (1–6) does not belong?

Brainteasers

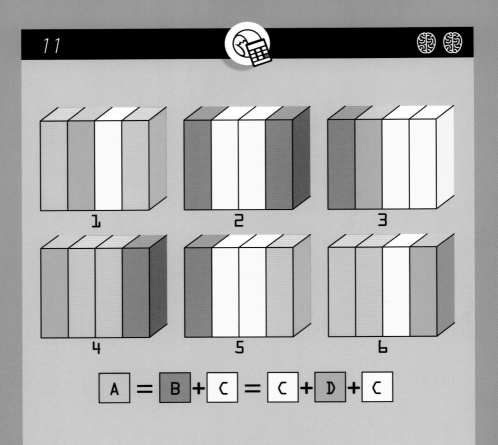

Which insulation (1–6) will insulate the best knowing that layer A insulates just as much as layer B+C or as layer C+D+C?

Brainteasers

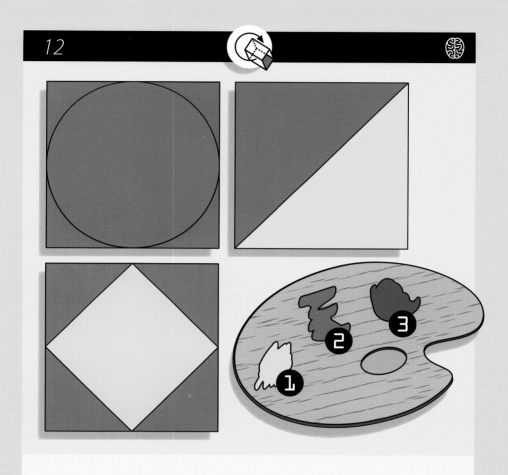

12

Which paint (1–3) was used the most to color in the three shapes?

Brainteasers

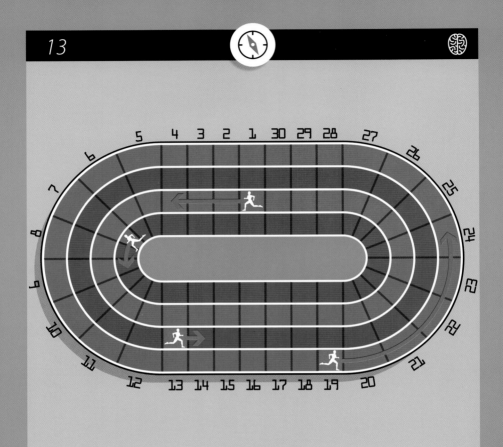

On which line (1–30) will all the runners be next to each other if every runner moves at his indicated speed (number of squares) per unit?

Brainteasers

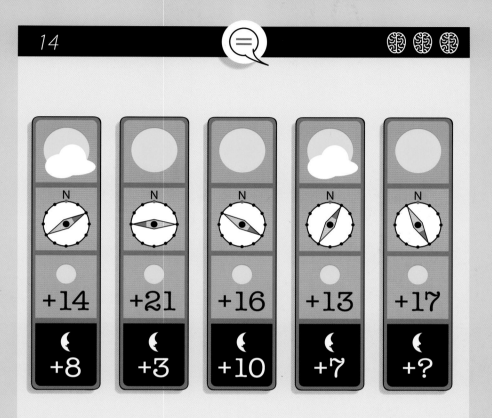

Knowing that there is a logical connection between the temperature and wind direction, which night temperature should replace the question mark?

Brainteasers

*The short-track race lasted 151 seconds.
Due to a defect in the digital clock, the hours, minutes,
and seconds of the time when the race began and
ended are only partly shown. In the upper right corner
you can see how the digits are displayed on a properly
functioning digital clock. When did the race end?*

Brainteasers

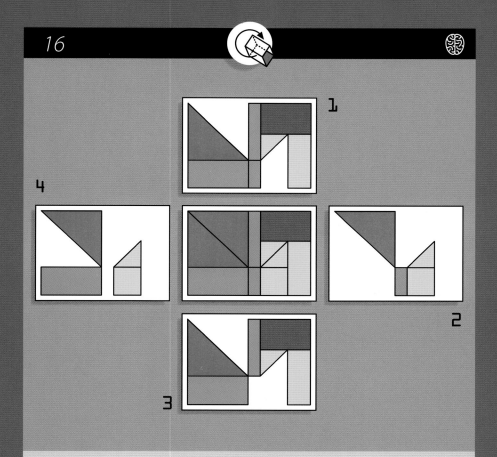

With which two groups of puzzle pieces (1–4) can you make the middle puzzle without any pieces left over?

Brainteasers

17

BrainSnack®

Lorem ipsum dolor <u>sit</u> amet, consectetuer adipiscing elit, sed diam nonummy nibh <u>duismoe</u> tincidunt ut laoreet <u>dolore</u> magna aliquam erat volutpat. Ut wisi enim ad <u>minin</u> veniam, quis nostrud exerci tation ullamcorper <u>suscipit</u> lobortis nisl ut aliquip ex ea commodo consequat. Duis auteb vel eum iriure dolor in hendrerit.

The underlined words each have two letters that satisfy a certain logic. What other word fits that same logic?

Brainteasers

18

A

B

C

D

*Knowing that their pedigree only includes
white and three other colors,
which butterfly (A–D) doesn't belong to the same family?*

Brainteasers

19

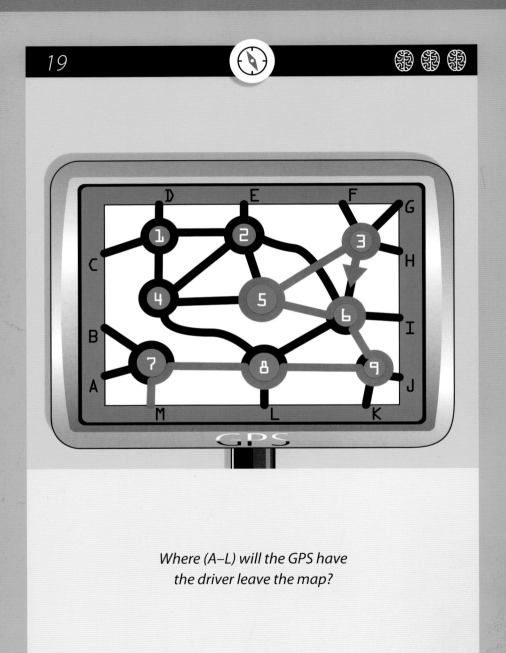

Where (A–L) will the GPS have the driver leave the map?

Brainteasers

20

1   2   3   4   5   6

Which rose (1–6) doesn't belong to
the same family as all the others?

31

Brainteasers

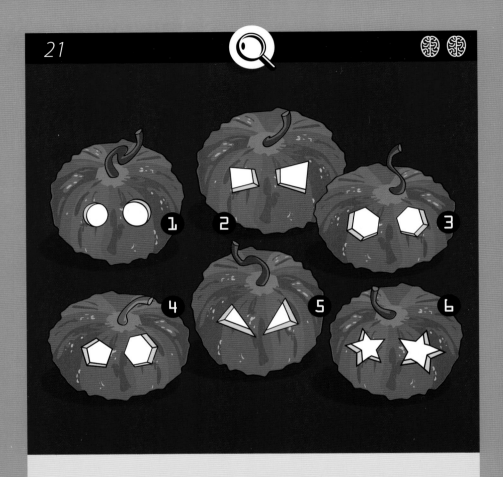

Which pumpkin (1–6) does not belong?

Brainteasers

22

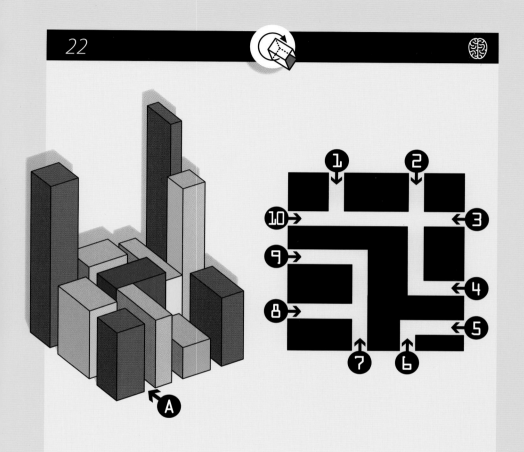

We enter the city at point A.
With which point (1–10) on the
floor plan does point A correspond?

Brainteasers

23

The mouse has already collected four chunks of cheese.
Which chunk of cheese (1–6) will it collect now?

Brainteasers

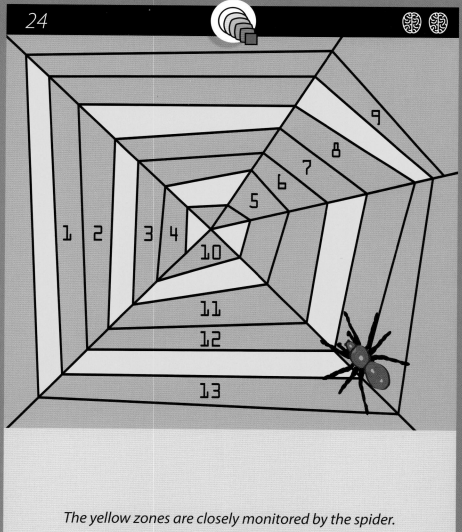

The yellow zones are closely monitored by the spider.
Which zone (1–13) should also be yellow?

Brainteasers

*Which group of clouds (1–6)
shows an impossible situation?*

Brainteasers

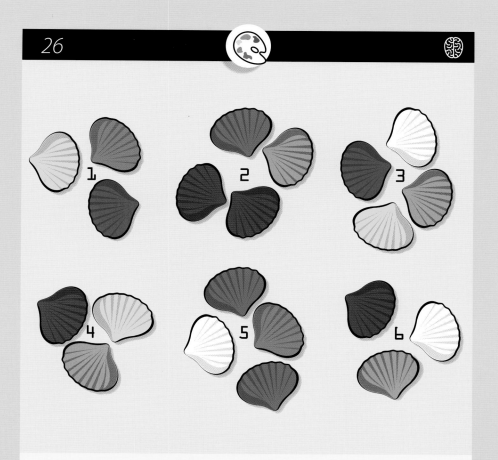

Which set of shells (1–6) does not belong?

Brainteasers

27

*Which three-letter word in the top
right-hand corner of the tickets is wrong?*

Brainteasers

28

How many people are in one skiff?

Brainteasers

29

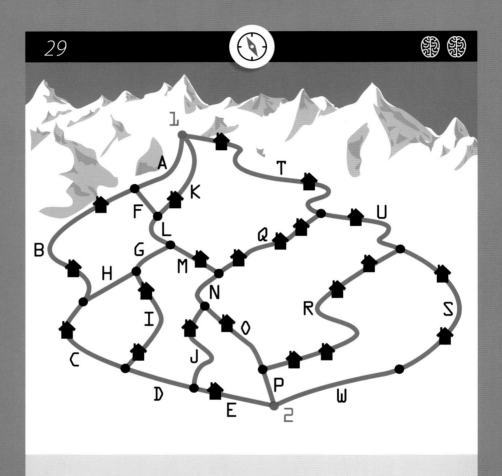

Find three separate ski runs that descend unremittingly from point 1 to point 2 and that do not cross each other at any time.

On which ski run do you encounter the most chalets? Answer with the letters (A–W) from top to bottom.

Brainteasers

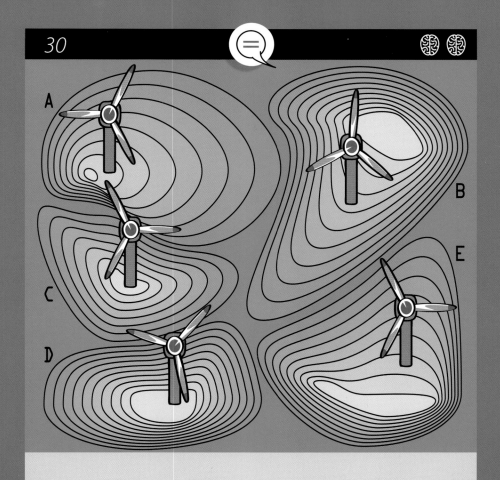

On which hill (A–E) is the windmill in the wrong place?

Brainteasers

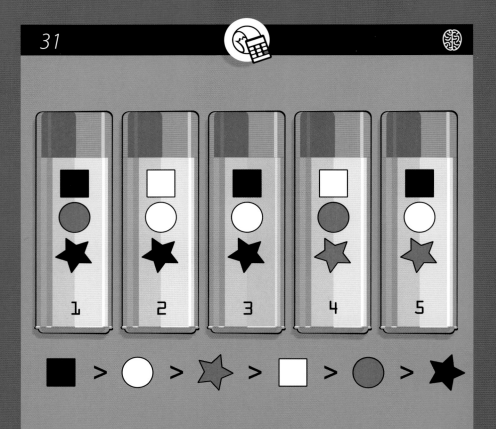

*Each spray can has three symbols
indicating the quality of the paint.
According to the table of comparison,
which spray can (1–5) contains the best paint?*

Brainteasers

32

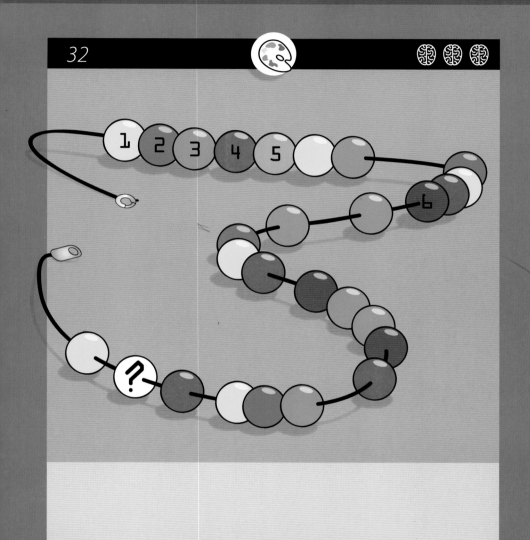

*Which color (1–6) should replace
the white bead with a question mark?*

Brainteasers

33

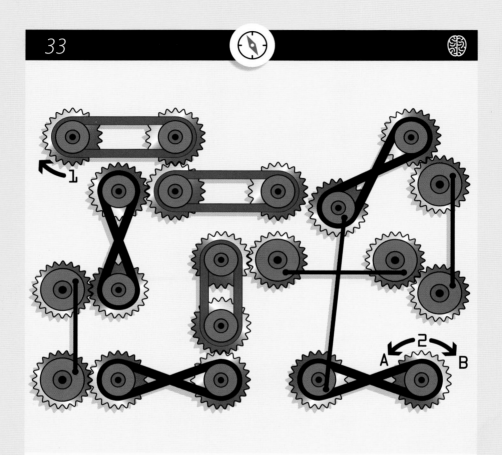

*Which direction (A or B) will element 2
turn if element 1 turns clockwise?*

Brainteasers

*What percent of fruit does the yogurt container
with the question mark contain?*

Brainteasers

35

19247 7429 924 42 ?

Brian drew up a diagram to drastically reduce the
amount of oil his transport company uses.
How many barrels does he still want to use?

Brainteasers

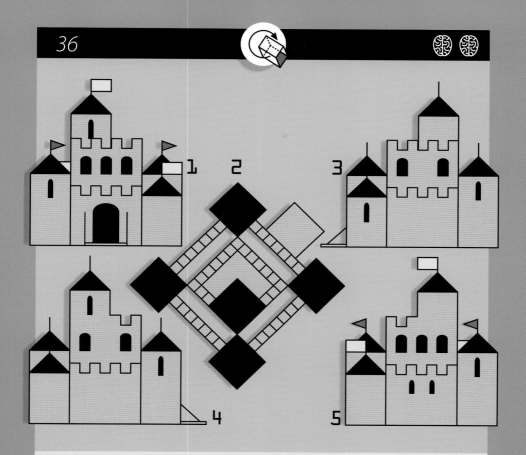

Which angle (1–5) of this castle is wrong?

Brainteasers

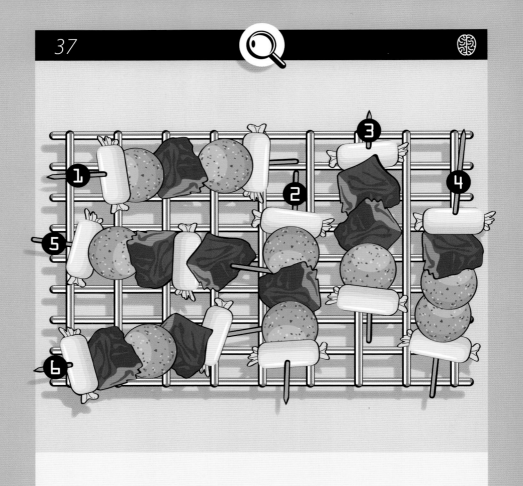

Which kebab (1–6) does not belong?

Brainteasers

38

Which rectangles (A–I) in this unfinished work of art still have to be colored in?

Brainteasers

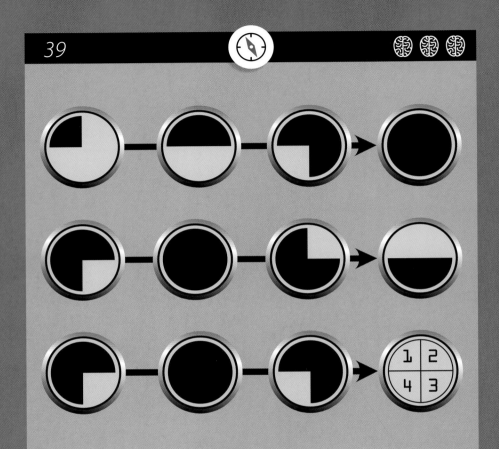

Which segments (1–4) have to be colored black in order to complete the series?

Brainteasers

40

1 2
3 4
5 6
7 8
9 10

*Which color (1–10) should
replace the question mark?*

Brainteasers

41

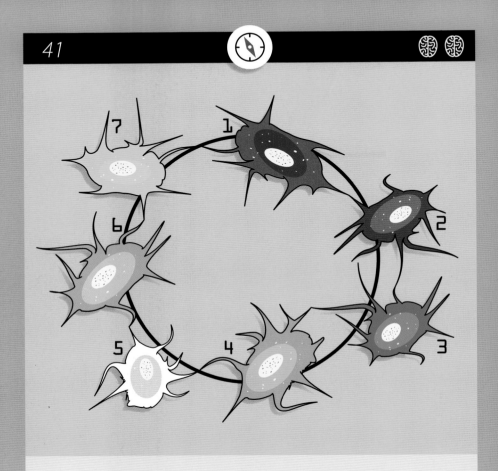

Which brain cell (1–7) is located five places farther clockwise from the brain cell that is located three places counterclockwise from the brain cell that is clockwise right next to the yellow brain cell (7)?

Brainteasers

How many items are in the TV-tube map?

Brainteasers

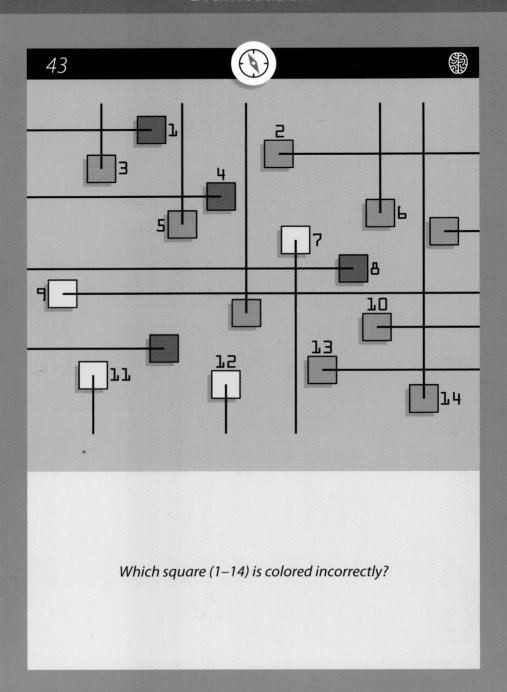

43

Which square (1–14) is colored incorrectly?

Brainteasers

44

*Which number should replace
the question mark on the cube?*

Brainteasers

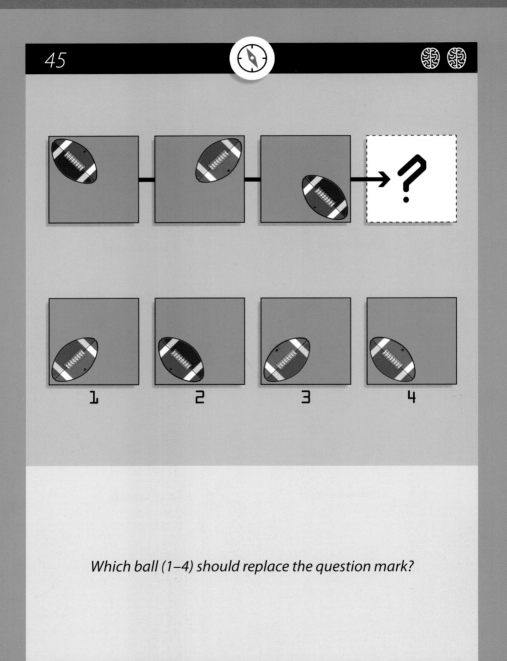

Which ball (1–4) should replace the question mark?

Brainteasers

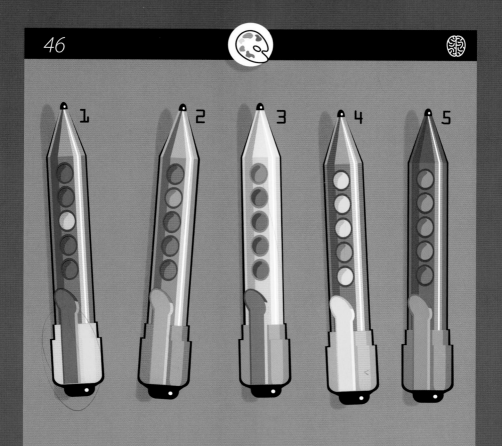

Which pen (1–5) with five colored dots doesn't belong in this set?

Brainteasers

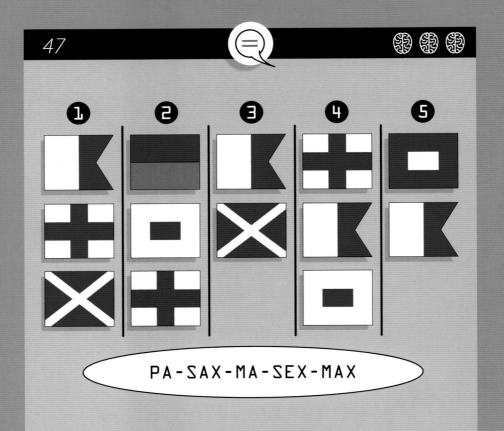

PA-SAX-MA-SEX-MAX

Each column contains the letters of a word in the
balloon below displayed with signal flags.
The letters in each column will be scrambled.
Enter the numbers of the columns in the same
order as the words in the balloon.
Answer like this: 31524.

Brainteasers

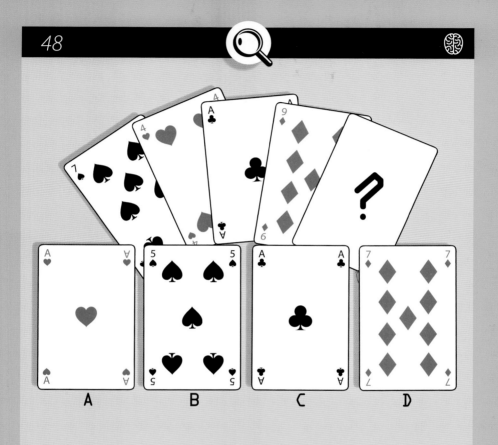

A   B   C   D

Which playing card (A–D) from the same deck is the only suitable replacement for the question mark?

Brainteasers

49

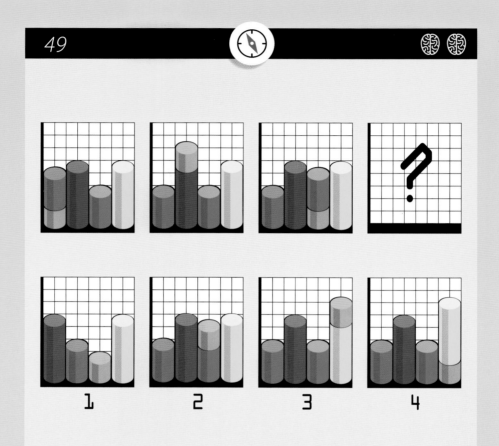

Which diagram (1–4) should replace the question mark?

Brainteasers

Which surfboard has the wrong contest number?
Answer like this: AD18S4.

Brainteasers

Which cars (1–6) belong in parking spaces A to G?
Answer like this: 1233556.

Brainteasers

52

The two groups of ice cubes form a cube of 3 by 3 blocks.
Which ice cube (1–9) should be removed?

Brainteasers

53

Which spinning movement (1–6) does not belong?

Brainteasers

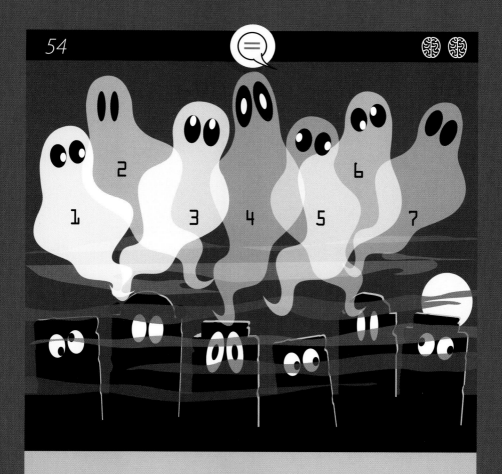

Which ghost (1–7) doesn't belong in this cemetery?

Brainteasers

55

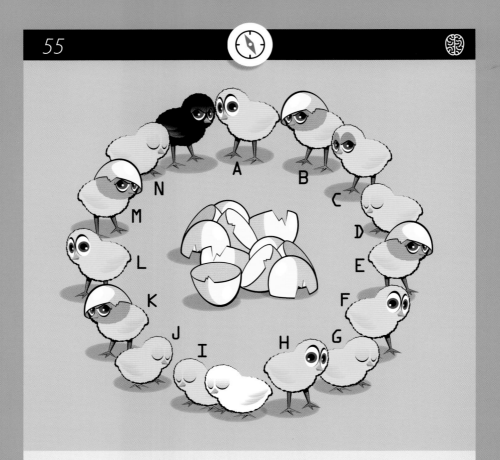

Which chick (A–N) is more than four chicks from the white chick and more than three chicks from the black one?

Brainteasers

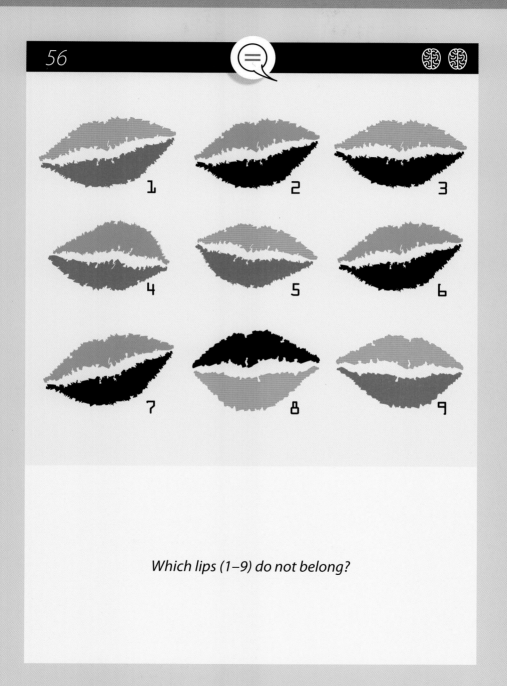

Which lips (1–9) do not belong?

Brainteasers

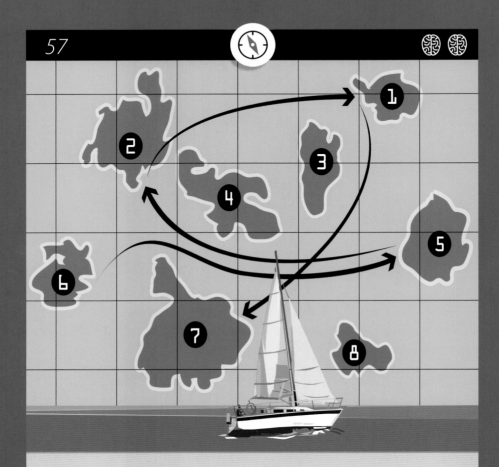

*In which order will the yacht
visit the three remaining islands?
Answer like this: 348.*

Brainteasers

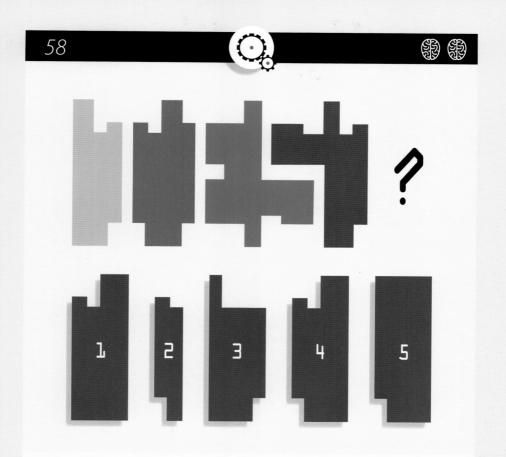

*Which patch (1–5) should
replace the question mark?*

Brainteasers

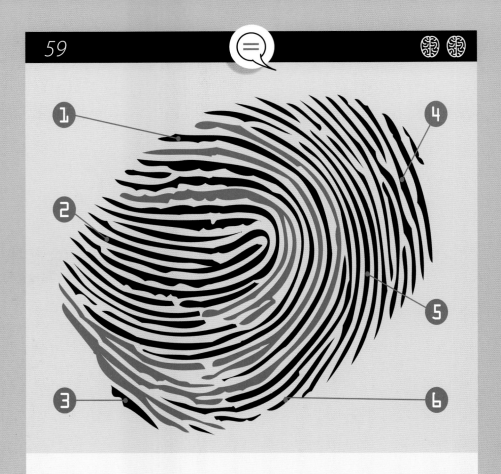

When investigating this fingerprint the
detective mainly relies on the red lines.
Which line (1–6) will he also use for identification?

Brainteasers

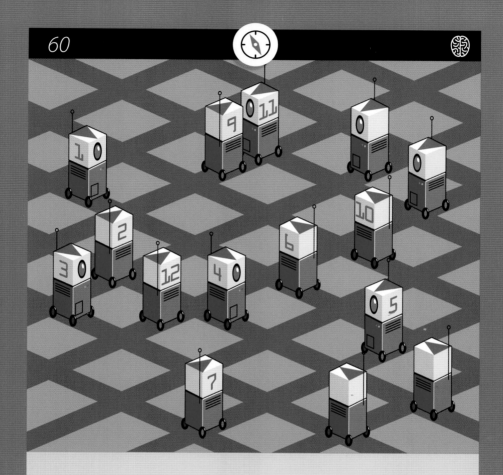

These inspection robots only work in pairs.
Which robot (1–12) isn't needed?

Brainteasers

61

Which two letters have to swap places so that the logic of the location of the letters is applied everywhere?

Brainteasers

62

How many different types
of baseball caps do you see?

Short-term Memory Games

Our short-term memory can contain an average of seven chunks of information.

Short-term memory

What is called "short-term" when talking about memory? A few seconds, a few minutes; it won't be much longer than that. Research has shown that in general the short-term memory only has room for seven chunks of information. Through concentration, emotions, repetition, associations, study aids, and tricks, these chunks in the short-term memory can be linked to our long-term memory. This enables us to remember more information better and longer. Naturally, the information that you think is important has a greater chance of moving to the long-term memory and not "disappearing" from your head. If you don't do this then temporarily saved information will vanish into thin air.

Remember, when playing the following memory games, try to link as much new information as possible to your long-term memory.

Tip!
Perhaps you should eat an extra banana because your brain uses about fifteen watts per day, which corresponds to the energy in two ripe bananas.

Short-term Memory Games

63

Because humans are by nature very visual, the average image span consists of nine elements. When the elements form a group, that will increase.

Look intently—for ninety seconds—at the circled products that you must purchase at the market. Afterward write down as many of the circled products as possible. Try to improve your score by viewing this photo again until you can remember them all.

Short-term Memory Games

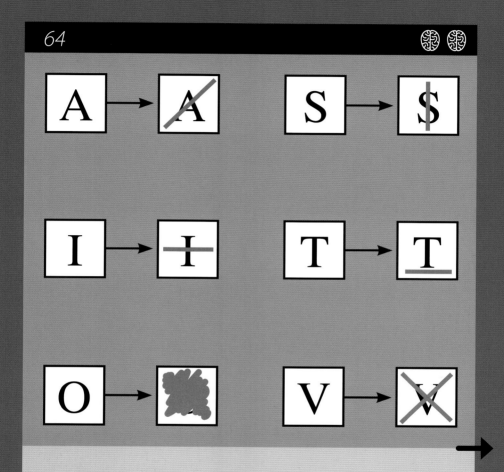

*Look intently at this image for sixty seconds and
remember how each letter is crossed off,
then complete the assignment on the back of this page.*

Short-term Memory Games

64 Continuation

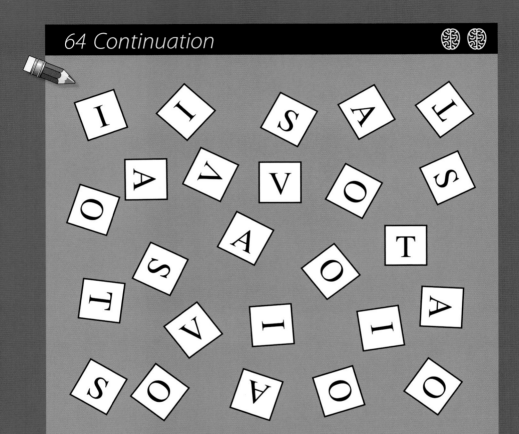

Try to cross off all the letters in the same way as on the previous page as quickly as possible.

Go back if necessary.

Short-term Memory Games

65

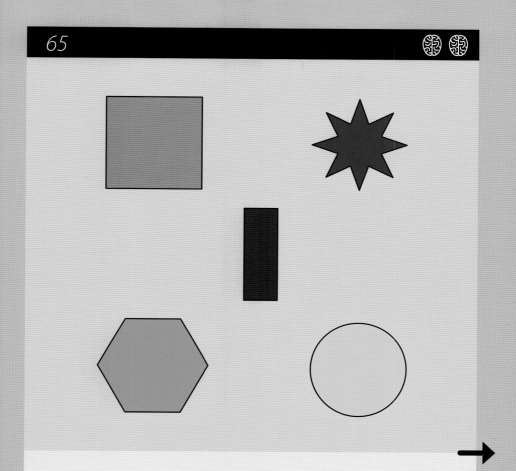

Look intently at the shapes, colors, and order for sixty seconds, then complete the assignment on the back of this page.

Short-term Memory Games

65 Continuation

1.
To the left of the numbers (1–5) mark the two shapes that have switched places.

2.
To the right of the numbers (1–5) mark the two shapes that have changed color.

Short-term Memory Games

66

Look at this black-and-white composition attentively for forty-five seconds, then complete the assignment on the back of this page.

Short-term Memory Games

66 Continuation

1.
Color the segments just like on the previous page.

2.
If necessary, go back and study the distance
between the colored segments.

Short-term Memory Games

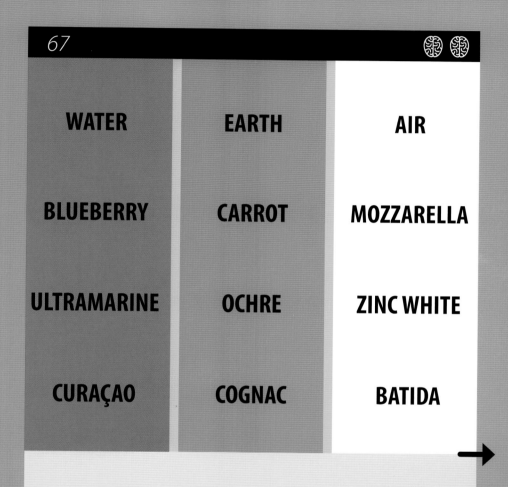

67

WATER	EARTH	AIR
BLUEBERRY	CARROT	MOZZARELLA
ULTRAMARINE	OCHRE	ZINC WHITE
CURAÇAO	COGNAC	BATIDA

*Look intently at these lists for sixty seconds
and find the relationships, then complete the
assignment on the back of this page.*

Short-term Memory Games

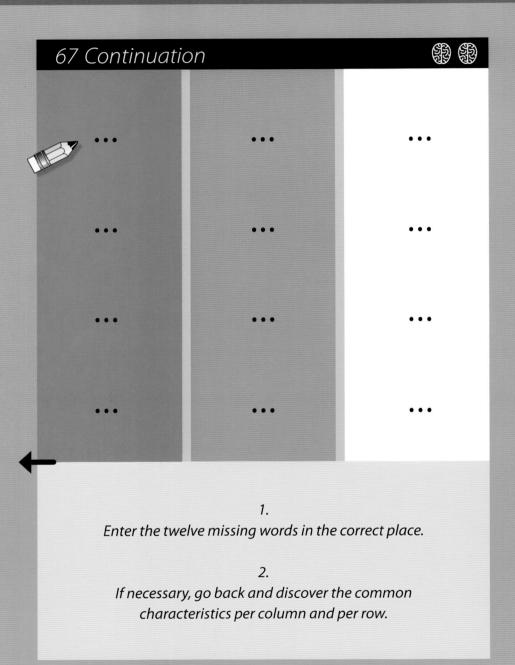

67 Continuation

1.
Enter the twelve missing words in the correct place.

2.
If necessary, go back and discover the common
characteristics per column and per row.

Short-term Memory Games

68

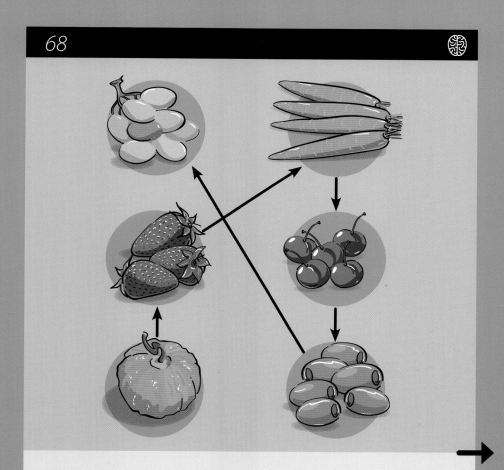

*Look intently at this healthy food
for forty-five seconds, then complete the
assignment on the back of this page.*

Short-term Memory Games

68 Continuation

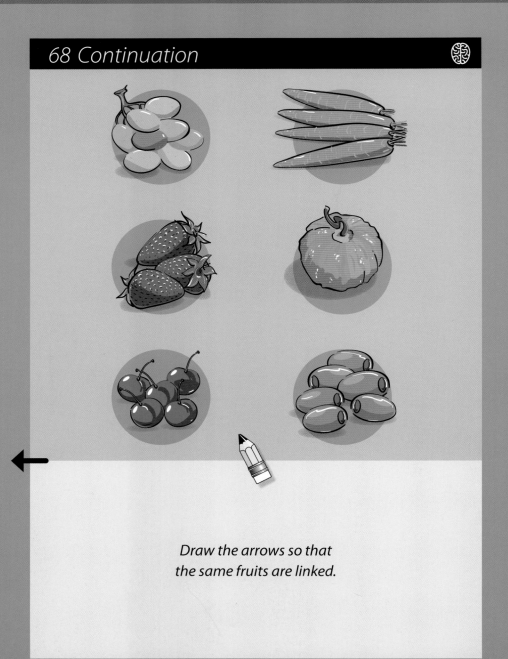

*Draw the arrows so that
the same fruits are linked.*

Short-term Memory Games

69

901

Penalty

Message

DECEPTION

1.09

ladder

→

Take your time and make up a story so that you remember the twelve elements (images, words, and numbers) below, then go to the next page and try to answer all the questions.

Short-term Memory Games

69 Continuation

1. *Recite your story again and write down*
    the twelve elements from the previous page.

    1 ....

    2 ....

    3 ....

    4 ....

    5 ....

    6 ....

    7 ....

    8 ....

    9 ....

    10 ....

    11..

    12 ....

2. *Which six elements are images?*

    1 ....

    2 ....

    3 ....

    4 ....

    5 ....

    6 ....

3. *Which three elements contain the color red?*

    1 ....

    2 ....

    3 ....

Short-term Memory Games

70

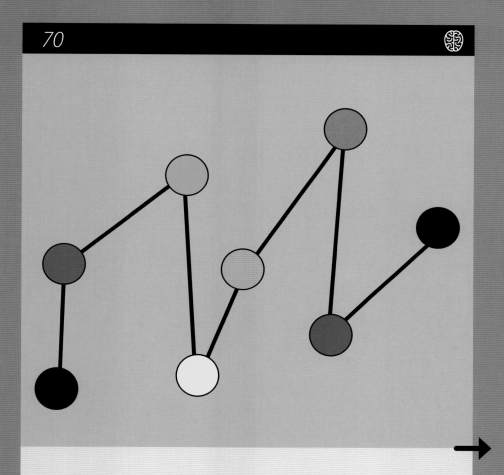

*Look intently at these colored dots for forty-five seconds.
Then complete the assignment on the back of this page.*

Short-term Memory Games

70 Continuation

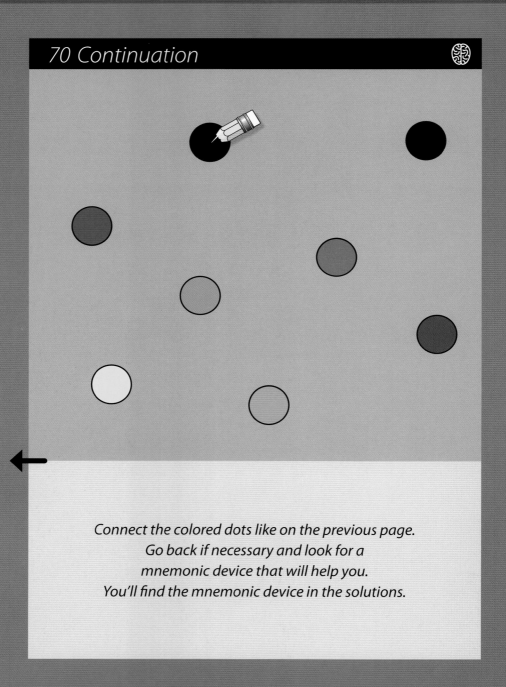

Connect the colored dots like on the previous page.
Go back if necessary and look for a
mnemonic device that will help you.
You'll find the mnemonic device in the solutions.

Short-term Memory Games

71

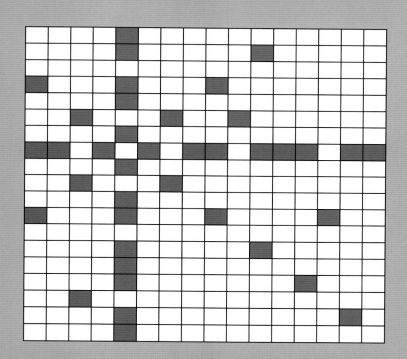

Look intently at the grid for sixty seconds,
then complete the assignment on the back of this page.

Short-term Memory Games

71 Continuation

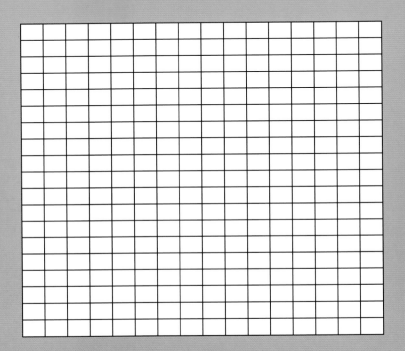

1.
Color the boxes just like on the previous page.

2.
*Go back if necessary and study
the distances between the boxes.*

Short-term Memory Games

72 KNOW YOUR NUMBER SPAN

A   4 8

B   3 9 2

C   6 0 1 5

D   4 8 9 2 0

E   5 6 0 7 9 2

F   9 7 3 5 1 7 6

G   0 8 4 5 9 2 7 4

H   2 7 1 0 8 4 5 2 3

I   1 9 7 6 4 5 9 4 3 2

J   7 4 9 6 5 3 0 4 9 5 1

K   8 4 3 0 7 9 4 7 2 0 6 9

L   5 4 8 7 6 2 0 1 5 4 8 9 7

M   4 5 4 8 7 0 2 3 6 5 2 1 9 6

1. Memorize the numbers in row A.
2. Write these numbers in the correct order on the back.
3. Repeat this for each row from B through M.
4. Check your answers afterward.

Your number span equals the quantity of numbers of the last row that you got correct. For example, if your first error was in row E with 6 numbers then your number span is 5.

Short-term Memory Games

72 Continuation

A . . .

B . . . .

C . . . . .

D . . . . . .

E . . . .

F . . . .

G . . . .

H . . . .

I . . . .

J . . . .

K . . . .

L . . . .

M . . . .

Write the series of numbers row after row here.

As numbers appear in a series of numbers that you can associate with a year of birth or a phone number, your score will be higher.

Short-term Memory Games

73

→

Look intently at these faces for sixty seconds,
then complete the assignment on the back of this page.

Short-term Memory Games

73 Continuation

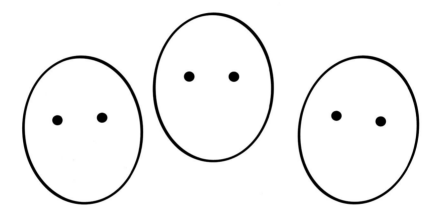

←

Complete the faces as shown on the back.
If you remember the series of numbers this task should work.

If necessary, go back and study the
numbers again for thirty seconds.

Short-term Memory Games

74

During pregnancy, fetuses make up to three thousand brain cells per second during certain periods. Once fully grown, the brain is composed of about a hundred billion neurons. Between each cell and hundreds of neighboring cells up to ten thousand synaptic connections can be made as we learn something new. These first weak connections become stronger upon repetition or when intense emotions influence the learning process.

Read this brain fact very carefully several times until you remember everything, then perform the assignment on the back of this page.

Short-term Memory Games

74 Continuation

During pregnancy, babies make up to three thousand brain cells per minute during certain periods. Once fully grown, the brain is composed of about a hundred million neurons. Between each cell and hundreds of neighboring neurons up to ten thousand synoptic connections can be made as we learn something new. These first weak connections become weaker upon repetition or when intense emotions influence the learning process.

←

Improve the six differences with the original text.

Short-term Memory Games

75

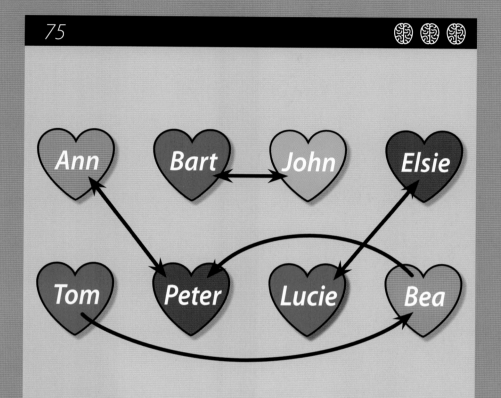

Carefully study who is in love with whom.
Hearts with the same color are brothers and sisters.

Once you have remembered everything,
complete the assignment on the back of this page.

Short-term Memory Games

75 Continuation

On each vertical line, color the hearts of who is in love with whom and mark who is brother and sister.

Short-term Memory Games

$$8+4= \quad 3+9=$$

$$2+10= \quad 17-5=$$

$$1+11= \quad 6+6=$$

*Look intently at the calculations for forty-five seconds,
then complete the assignment on the back of this page.*

Short-term Memory Games

76 Continuation

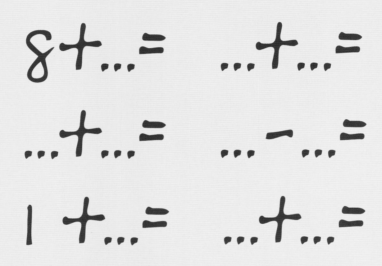

Enter all the missing numbers.

Short-term Memory Games

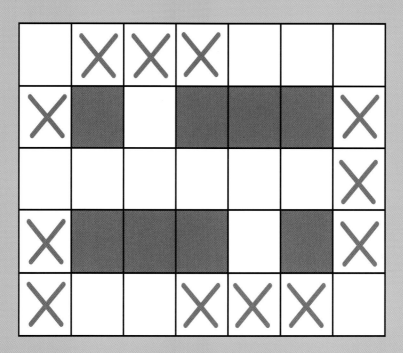

*Look intently at this grid for sixty seconds,
then complete the assignment on the back of this page.*

Short-term Memory Games

77 Continuation

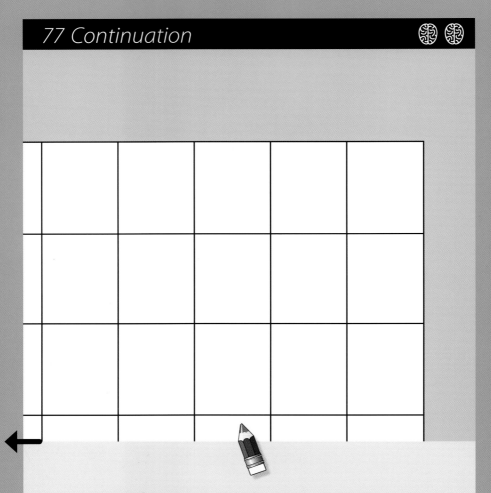

Color and place crosses in the fifteen squares so that this grid is identical to the upper right-hand corner of the grid on the back.

If necessary, go back and study
the squares again for thirty seconds.

Short-term Memory Games

78

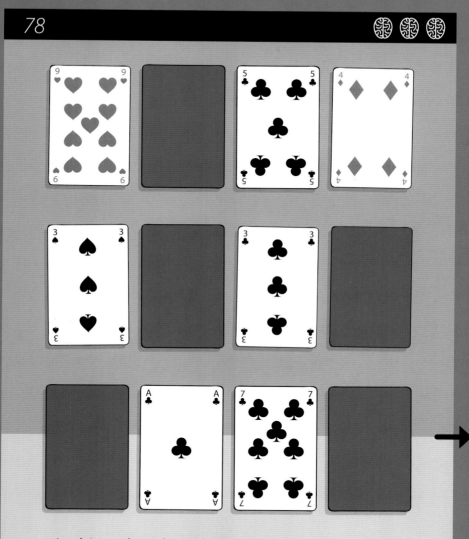

Look intently at these playing cards for ninety seconds, then complete the assignment on the back of this page.

Short-term Memory Games

78 Continuation

Just like in the game Memory, there are six pairs hidden in the twelve cards whose value and color are identical. You already see two pairs. The cards were not moved. Search for the other pairs.

Short-term Memory Games

79

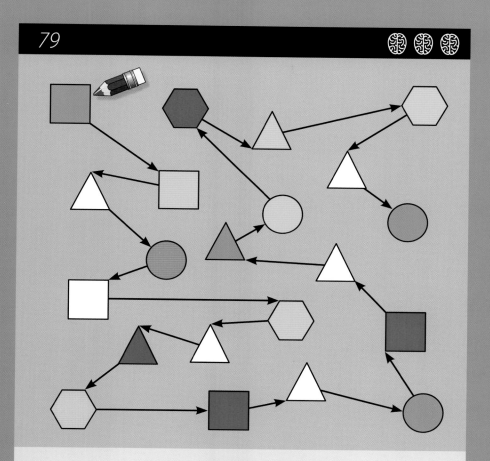

Spoil your short-term memory.

1. Mark the four shapes as quickly as possible just like in the key below following the direction of the arrows.
2. Leave the shapes that do not appear in the key untouched.
3. The number in a white triangle increases each time by one.
4. Try to look at the key as little as possible.

109

Figure It Out!

$a = 9.99999\ldots$

$10a = 99.99999\ldots$

$10a - a = 90$

$9a = 90$

$a = 10$

So $10 = 9.99999\ldots$

Is this possible?

Solution on page 288

0 6

3

1

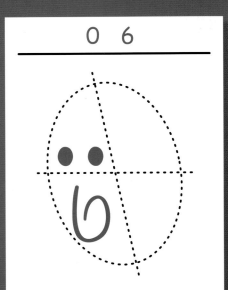

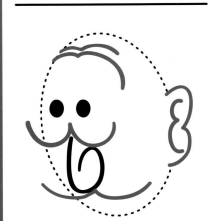

Figure It Out!

Number mania

Look, a wizard at mathematics, a mathematical genius! That's how we popularly term someone who has "talent" or a feeling for numbers and figures. Some say this "feeling for" is innate. Others disagree; even at a later age you can learn to calculate. Practice often and, if possible, practice in a playful manner.

With a fun game you'll learn to solve arithmetic problems and form series of numbers whose mathematical beauty you'll learn to appreciate afterward. By playing such number games often, you'll notice that it becomes easier to do sums in daily life.

Tip!
Here's a fun calculating and memory exercise to know whether you paid too much at the register and whether you have exceeded your budget: Add the price of a product that you are going to buy to the sum of the products that are already in your cart. This is a fun exercise for children, too.

Believe it or not, you'll have lots of fun and it's beneficial, too.

Figure It Out!

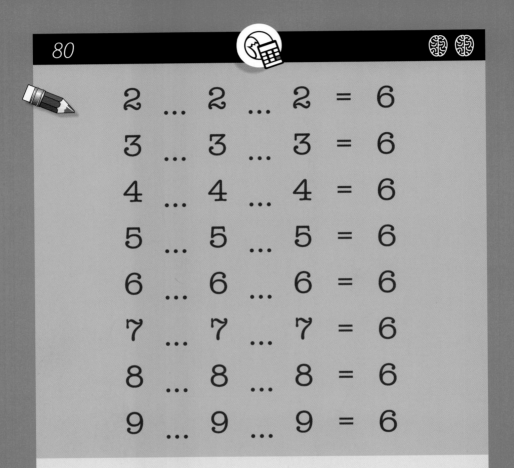

80

$$2 \ldots 2 \ldots 2 = 6$$
$$3 \ldots 3 \ldots 3 = 6$$
$$4 \ldots 4 \ldots 4 = 6$$
$$5 \ldots 5 \ldots 5 = 6$$
$$6 \ldots 6 \ldots 6 = 6$$
$$7 \ldots 7 \ldots 7 = 6$$
$$8 \ldots 8 \ldots 8 = 6$$
$$9 \ldots 9 \ldots 9 = 6$$

Place the correct operations (addition, subtraction, multiplication, division, square root) next to the numbers above so that the solution always equals 6.

Figure It Out!

81

A farmer keeps his chickens and pigs in the same pen.
Between them there are 72 heads and 200 legs.
How many of each animal does the farmer have?

Figure It Out!

82

Fill in this magic triangle with the numbers from 1 to 9 so that the sum of the numbers on each side equals 20.

Take into account the < (less than) and > (greater than) signs between the numbers.

Figure It Out!

9   7

14   18

27   21

28   3?

*Which digit should replace the
question mark in the score?*

Figure It Out!

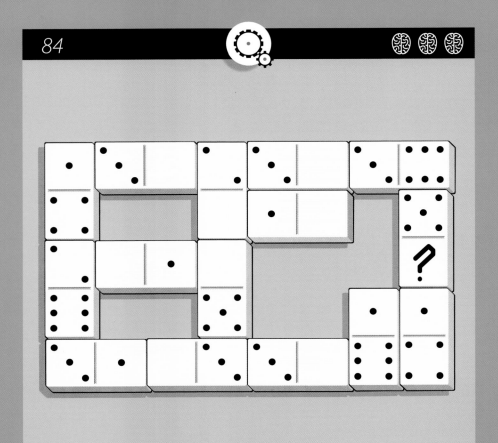

*How many spots should replace the
question mark on the domino?*

Figure It Out!

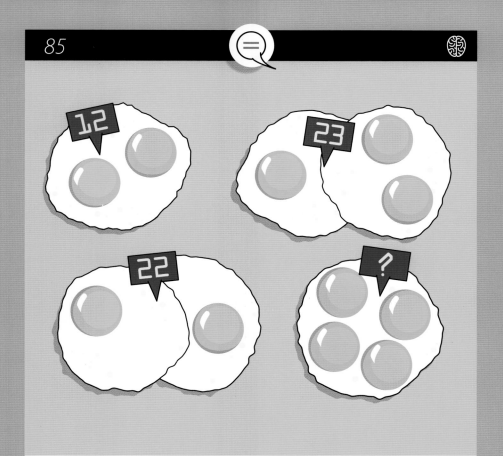

Which number is missing on the fried egg with four yolks?

Figure It Out!

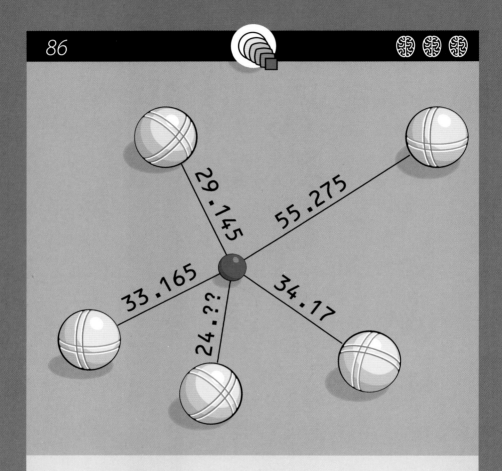

86

Which number should replace the question marks?

Figure It Out!

87

On August 18 a yacht with a draught of six feet wants to enter a marina. To ensure that the yacht is not damaged, the skipper wants a minimum of three feet free depth under the keel.

From the tidal table, which was drawn up in wintertime, he can calculate that the water depth at 8:00 A.M. is five feet and that it increases by one foot per hour. As of what time can he enter the marina?

Figure It Out!

Which number should replace the question mark?

Figure It Out!

0000 = 0   0100 = 4
0001 = 1   0101 = 5
0010 = 2   0110 = 6
0011 = 3   0111 = 7

? = 12

*The table shows the binary notation
of the decimal numbers from 0 to 7.
What is the binary notation of the decimal number 12?*

Figure It Out!

Which number should replace the question mark?

Figure It Out!

```
    Market
    75.00
     ?.??
     3.18
     1.82
   ─────────
    8?.??
```

The price of the second fish on this receipt is missing.

Find one single amount that consists of three digits and can indicate both the cost of the fish and the missing digits in the total price of all the fish.

Figure It Out!

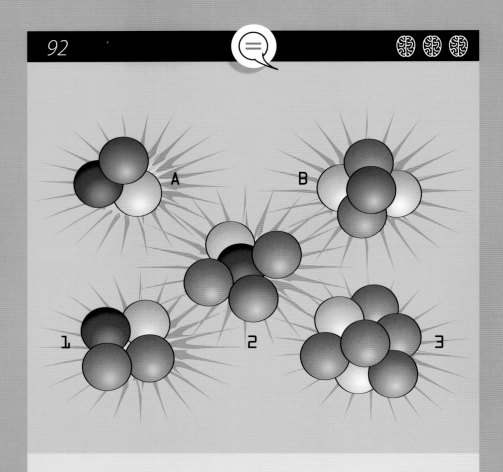

Knowing that nucleus A has just as much energy as nucleus B, which two nuclei (1–3) have the same energy value?

Figure It Out!

If one and a half boys can eat one and a half hamburgers in one and a half minutes, how many hamburgers can six boys eat in six minutes?

Figure It Out!

Knowing that the score of archer A is 55 and that different points are earned in the ten zones in ascending value toward the center, how many points does archer B score?

Figure It Out!

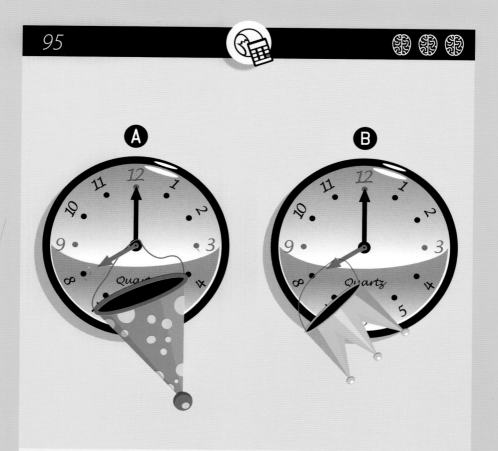

The party starts at both places at 8 P.M.
Clock A runs twenty minutes behind every hour and
clock B runs fifteen minutes behind every hour.

The party will end as soon as the big hand points to twelve
on both clocks. After how many hours will the party end?

Figure It Out!

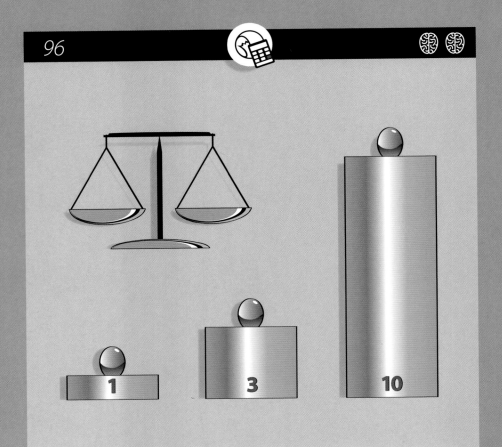

Which weight between 1 and 14 ounces cannot be weighed with the three available weights in only one weighing?

Figure It Out!

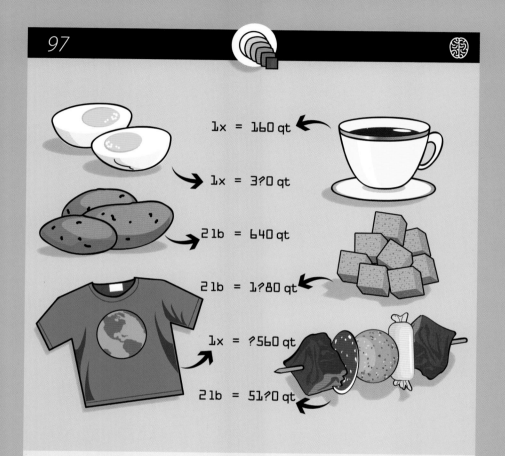

97

1x = 160 qt

1x = 3?0 qt

2 lb = 640 qt

2 lb = 1?80 qt

1x = ?560 qt

2 lb = 51?0 qt

Next to each product you see the average number of quarts of water needed to produce the indicated amount. To produce one cup of coffee you need approximately 160 quarts of water. Which unique number is missing four times?

Figure It Out!

```
    1 2 ?
 + 7 ? 8 9 4
 − 3 5 6 ? 4
 _____
   ? ? ? ? ?
```

Which three different numbers are missing above the line to make the number below the line as big as possible?

Figure It Out!

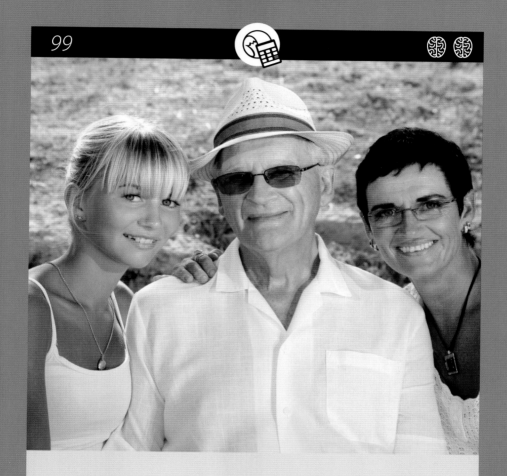

99

Peter, Ella, and Linda are collectively 120 years old. Linda is three times older than Ella who is six times younger than Peter. List each person's age.

Figure It Out!

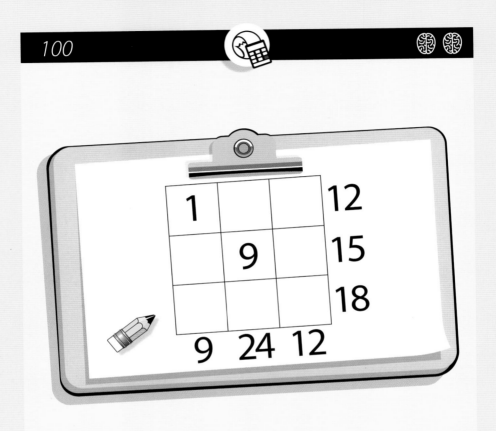

Fill in the chart with all the numbers from 1 to 9 so that the sum of the numbers per row and per column corresponds with the numbers on the side. Answer with the 9 numbers from left to right.

Figure It Out!

101

A

5	21	8
16	13	1
9	●	7
6	20	22
12	10	11
2	27	17
23	18	●
14	19	24
●	25	3

B

5	21	●
16	13	1
9	26	7
6	20	22
12	10	11
2	27	17
23	18	4
14	●	24
15	25	3

C

5	21	8
16	●	1
9	26	7
6	●	22
12	10	11
●	27	17
23	18	4
14	19	●
15	25	3

D

5	21	8
16	13	1
9	26	●
6	20	22
12	10	11
2	27	17
23	18	4
14	19	24
15	25	3

On which number should the token be placed in section D?

Figure It Out!

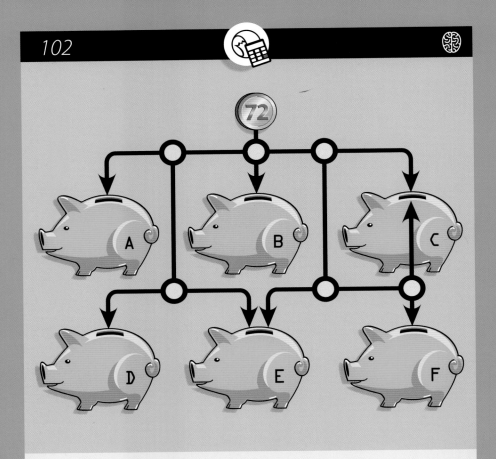

Which piggy bank (A–F) will receive the largest
amount of the seventy-two dollars knowing that the
money is divided proportionally when split up?

Figure It Out!

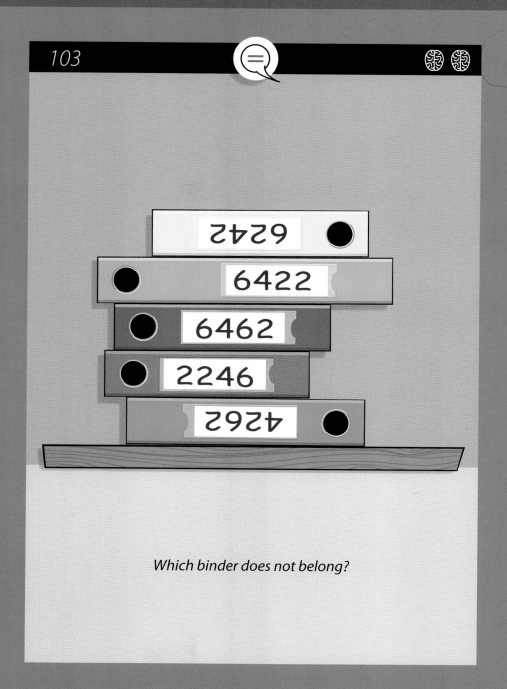

Which binder does not belong?

Figure It Out!

19 11 11 ? 12

Which number should replace the question mark?

Figure It Out!

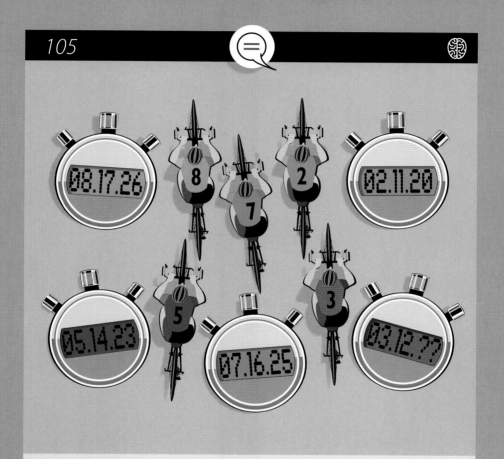

How many seconds should replace the question marks on the chronometer of rider 3?

K?w?ge
&
l?gu?e

Reaserch has sohwn taht the
odrer of the lrettes in a wrod
is not imorptant if the frist
and lsat letetrs are in palce.
Apaprnetly the bairn sees the
wrod as a uint and not lkie
speratae leterts.

Long-term memory

A comparison often provides clarity. The "working memory" of a computer is best compared to your own short-term memory (everything that you remember briefly) and the hard disk is like your long-term memory, all impressions and information saved upstairs that you can call up at any time to solve new puzzles and problems.

Pure word problems call upon all the data and knowledge in your long-term memory. You test new material against existing information via associations and logical connections. That way you make progress and think of adequate solutions.

Combinations that feel logical will be remembered longer and will be used more effectively when new word problems come up. It is a game; a relaxing game that you cannot get enough of.

Detail!
The speed that new and existing information moves between your brain cells to solve a puzzle ranges from 1 to more than 268 miles/hr (120 meters per second).

Knowledge & Language

106 EVOLUTION QUIZ

1 The first living creatures originated 3.8 billion years ago, but evolutionary development only exploded one billion years ago. Order the following forms of life according to their chronological development: amphibians, reptiles, fish, and mammals.

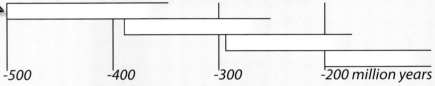

-500   -400   -300   -200 million years

2 In 1856 archaeologists found human remains in Neanderthal in Germany. Named after this valley, Neanderthals spread across Europe, the Middle East, and Central Asia between 150,000 and 28,000 years ago. Mark the three boxes that are applicable to Neanderthals.

☐ 1 His brains were bigger than those of modern man.
☐ 2 He made the transition to North America.
☐ 3 He used woven textile.
☐ 4 He lived during the last ice age.
☐ 5 He is the ancestor of current human beings.
☐ 6 He didn't eat meat.
☐ 7 He farmed the land.
☐ 8 He buried his own sort.

3 Einstein died in his sleep on April 18, 1955. He is the father of the theory of general relativity. These are a few of his sayings: "God does not throw dice," "Logic gets you from A to B, imagination takes you everywhere," "The search for truth is more precious than its possession," and "Fantasy is more important than knowledge, because knowledge is restricted."
What does the c stand for in his famous formula $E = mc^2$?

☐ 1 gravity   ☐ 2 speed of light   ☐ 3 electric charge

Knowledge & Language

107 LETTER BLOCKS

1 Numbers

2 Weather phenomena

3 Occupations

*Move the letter blocks around so that
words are formed on top and below that
you can associate with the subject.*

Knowledge & Language

108 TRUE OR FALSE?

1 Did you know...
The weight of your brain is only 3 percent of your total body weight.

4 Did you know...
In the beginning of pregnancy the fetus forms about three hundred neurons per second.

2 Did you know...
The brain uses about fifteen watts per day, which corresponds with the energy of two ripe bananas. It uses a constant amount of energy. Deep thought doesn't increase the energy consumption much.

5 Did you know...
The majority of the brain cells are already formed at birth. Growth stops around eighteen years old.

3 Did you know...
Approximately fourteen pints of blood flow through the brain per minute.

6 Did you know...
The brain uses about 20 percent of all oxygen that is pumped through the body.

Which surprising brain facts are true and which are false?

	TRUE	FALSE
1	☐	☐
2	☐	☐
3	☐	☐
4	☐	☐
5	☐	☐
6	☐	☐

Knowledge & Language

109 TRUE OR FALSE?

1   Did you know...
The brain is 3/4 water.

2   Did you know...
Brains cannot feel pain, in contrast to the blood vessels in the brain. Headaches are often a consequence of constricting and dilating blood vessels.

3   Did you know...
The subconscious starts working about one minute after the blood supply to the brain is stopped.

4   Did you know...
The speed at which information moves between the neurons ranges from 0.3 to more than 74 miles/hour.

5   Did you know...
The gray (pink) matter in the brain consists of the cell bodies of brain cells. The white matter consists of axons that are surrounded by a fatty insulating layer of myelin, which is responsible for the white color.

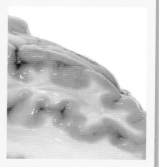

Which surprising brain facts are true and which are false?

	TRUE	FALSE
1	☐	☐
2	☐	☐
3	☐	☐
4	☐	☐
5	☐	☐

Knowledge & Language

110 LANGUAGE EXERCISE

Abort	_ _ _ _ _ _	Neater	_ _ _ _ _ _
Ballet	_ _ _ _ _ _	Open	_ _ _ _ _ _
Cake	_ _ _ _ _ _	Plugged	_ _ _ _ _ _
Debate	_ _ _ _ _ _	Queen	_ _ _ _ _ _
Easter	_ _ _ _ _ _	Rack	_ _ _ _ _ _
Fall	_ _ _ _ _ _	Sable	_ _ _ _ _ _
Graceful	_ _ _ _ _ _	Taken	_ _ _ _ _ _
Hello	_ _ _ _ _ _	Unite	_ _ _ _ _ _
Indemnify	_ _ _ _ _ _	Veal	_ _ _ _ _ _
Jaws	_ _ _ _ _ _	White	_ _ _ _ _ _
Kneel	_ _ _ _ _ _	x-axis	_ _ _ _ _ _
Layer	_ _ _ _ _ _	Yacked	_ _ _ _ _ _
Model	_ _ _ _ _ _	Zeroes	_ _ _ _ _ _

Replace one letter in each word so that you obtain a different word.

Several solutions are possible.

Knowledge & Language

111 GEOMETRY

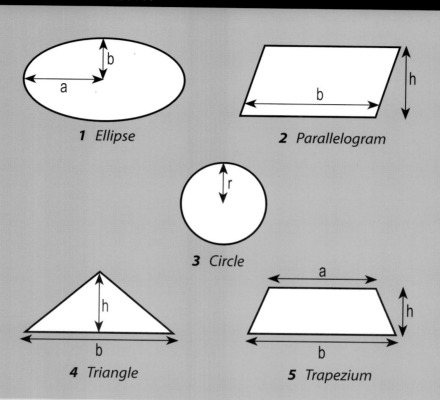

1 Ellipse

2 Parallelogram

3 Circle

4 Triangle

5 Trapezium

How does one calculate the surface of shapes?
It was once everyday fare, but practice makes perfect.

List the surface formulas of the shapes above.

Knowledge & Language

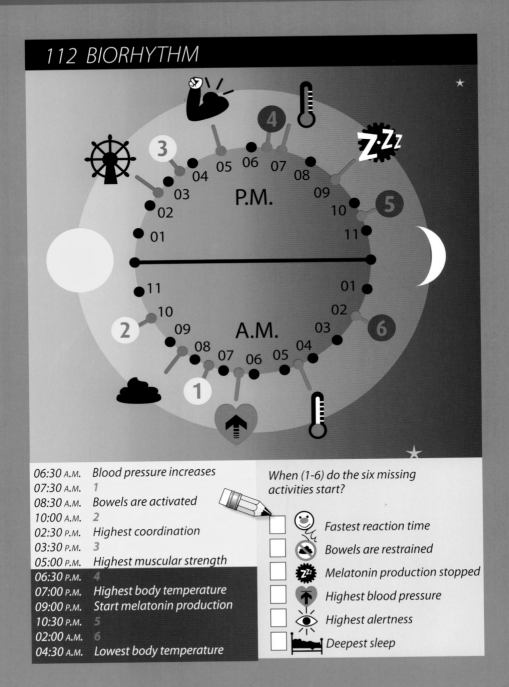

112 BIORHYTHM

06:30 A.M.	Blood pressure increases
07:30 A.M.	1
08:30 A.M.	Bowels are activated
10:00 A.M.	2
02:30 P.M.	Highest coordination
03:30 P.M.	3
05:00 P.M.	Highest muscular strength
06:30 P.M.	4
07:00 P.M.	Highest body temperature
09:00 P.M.	Start melatonin production
10:30 P.M.	5
02:00 A.M.	6
04:30 A.M.	Lowest body temperature

When (1-6) do the six missing activities start?

- [ ] Fastest reaction time
- [ ] Bowels are restrained
- [ ] Melatonin production stopped
- [ ] Highest blood pressure
- [ ] Highest alertness
- [ ] Deepest sleep

Knowledge & Language

113 ON THE DOPE LIST

1 Declaration (used to stimulate muscle and bone growth)

| | | | B | | | I | | S | | | | O | | |

2 Citron (taken for pleasure or to reduce pain)

| | A | | C | | | | | S |

3 Bracelet (decreases force and rate of heart contractions)

| B | | | | | | | O | | K | | |

4 Morons (secretion of an endocrine gland)

| H | | | | | E | |

5 Sunni Li (accelerates oxidation of sugar in cells)

| | | | | | | |

*Form the word that is described between
parentheses with the letters above each grid.
One or more extra letters are already in the right place.*

Knowledge & Language

114 3xLINK

How does one play 3xLink?

Connect the words that belong together in the left and right column. In the right column there is one word too many that must not be used.

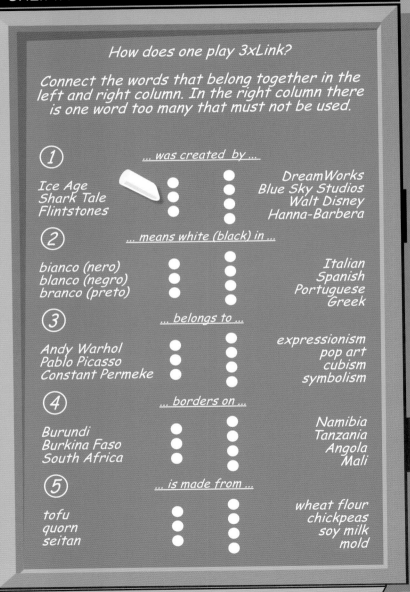

(1) ... was created by ...

Ice Age	DreamWorks
Shark Tale	Blue Sky Studios
Flintstones	Walt Disney
	Hanna-Barbera

(2) ... means white (black) in ...

bianco (nero)	Italian
blanco (negro)	Spanish
branco (preto)	Portuguese
	Greek

(3) ... belongs to ...

Andy Warhol	expressionism
Pablo Picasso	pop art
Constant Permeke	cubism
	symbolism

(4) ... borders on ...

Burundi	Namibia
Burkina Faso	Tanzania
South Africa	Angola
	Mali

(5) ... is made from ...

tofu	wheat flour
quorn	chickpeas
seitan	soy milk
	mold

Knowledge & Language

115 LETTER BLOCKS

1 *Extremes*

2 *Toys*

3 *Kitchenware*

Move the letter blocks around so that words are formed on top and below that you can associate with the subject.

The letters were reversed on one block.

Knowledge & Language

116 TRIVIA

1

How do you know if an egg is fresh if you place it in water?

- A it sinks
- B it floats
- C it spins around

2

What is the distance of a marathon?

- A 26 miles and 494 yards 42.295 km
- B 26 miles and 385 yards 42.195 km
- C 26 miles and 275 yards 42.095 km

3

Coconuts grow on which tree?

- A a date palm
- B a palm tree
- C a baobab

4

What is the name of the largest ocean in the world?

- A the Atlantic Ocean
- B the Indian Ocean
- C the Pacific Ocean

5

What is an antidote?

- A a virus
- B a vaccine
- C an antitoxin

Knowledge & Language

117 WORD ASSOCIATIONS

1 e.g. Apple - Eve = Adam

2 Crime - hammer =   ...

3 Hand - boss =   ...

4 Ocean - sport =   ...

5 Fruit - talking =   ...

6 Tennis - religion =   ...

7 Food - Ping-Pong =   ...

8 Brain - hard disk =   ...

9 Butterfly - tank =   ...

10 Tomato - balls =   ...

11 Film - board =   ...

12 Sun - smart =   ...

13 Tropical - coniferous tree = ...

14 Coffee - computer =   ...

Look for the element of accord between the two words.
This type of exercise results in lots of associations.

Several solutions are possible.

Knowledge & Language

118 DOODLE PUZZLE

SP
___
SOR

CLOUD
___
TH

Solution:

SPONSOR
SP *on* **SOR**

1 ....

2 ....

WAV

HER

3 ....

4 ....

5 ....

A doodle puzzle is a combination of images, letters, and numbers that indicate a word or a concept. If you cannot solve a doodle puzzle, do not look at the answer right away. Try to solve it later or tomorrow. When you know the answer, study the exercise and the solution to remember the structure and connections of the puzzle forever. Afterward it will be easy to solve doodle puzzles and explain them to your friends. This will reinforce your comprehension.

Knowledge & Language

119 EUPHEMISMS

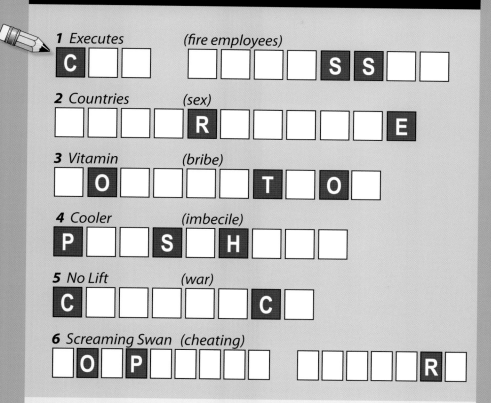

1 Executes (fire employees)

C [ ] [ ]   [ ] [ ] [ ] [ ] S S [ ] [ ]

2 Countries (sex)

[ ] [ ] [ ] R [ ] [ ] [ ] [ ] E

3 Vitamin (bribe)

[ ] O [ ] [ ] [ ] T [ ] O [ ]

4 Cooler (imbecile)

P [ ] [ ] S [ ] H [ ] [ ] [ ]

5 No Lift (war)

C [ ] [ ] [ ] [ ] C [ ]

6 Screaming Swan (cheating)

[ ] O [ ] P [ ] [ ] [ ] [ ]   [ ] [ ] [ ] [ ] R [ ]

A euphemism is a way of saying something more beautifully than it actually is, like penitentiary officer for jailer and buxom for fat.

Form the euphemism for the word that is described between parentheses with the letters above each grid. One or more extra letters are already in the right place.

Knowledge & Language

120 CHRONOLOGY

*Number the (international) events
or days that we commemorate in
the correct chronological order.*

Mother's Day

World Food Day

Super Bowl Sunday

International Women's Day

St. Patrick's Day

Winter Time (End of DST)

World Environment Day

Columbus Day

Father's Day

Christmas

New Year's Day

Easter Day

Labor Day

Martin Luther King Day

Valentine's Day

Memorial Day

Halloween

World AIDS Day

Knowledge & Language

121 LETTER BLOCKS

1 Shapes

2 American cars

3 Family

Move the letter blocks around so that words are formed on top and below that you can associate with the subject.

The letters were reversed on two blocks.

Concentration Exercises

Look before you leap

Concentration is an intense form of attention. The ability to concentrate is very valuable. For many people it is not easy to focus on something for a long time and to remain attentive. When you are focused you do not waste energy on unnecessary things.

Making associations, making mental notes, summoning pure concentration . . . it is all essential for remembering certain facts for a short or long period of time. Looking for associations and creating mental notes will only result in frustration without concentration.

Paying attention plays a crucial role in what you do. You do not have to think about every action, but you do have to remain attentive to the goal of your actions, for games, too, otherwise things will go wrong and you won't be able to solve the puzzle or brainteaser (quickly).

You can practice skills that you have not mastered. Part of your degree of concentration is genetically determined, but you can improve it. Attention training and concentration exercises can bring your intelligence and your ability to solve problems to a higher level.

Concentration Exercises

122 COUNTING LETTERS

Mentally calculate the sum of all the consonants of the nine colors.

Concentration Exercises

123 MENTAL ARITHMETIC

How much is 1/4 of 1/2 of 2^3 divided by 1/2 of the result?

Mentally calculate the above.

Concentration Exercises

124 MIRAGE

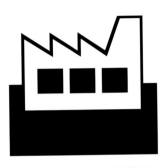

*Draw the two pictures as if
they were reflected in the water.*

Concentration Exercises

125 LEARN TO ESTIMATE

How high is your kitchen table?

ESTIMATION   PROOF

How high is this page?

ESTIMATION   PROOF

What is the size of your waist?

ESTIMATION   PROOF

Weigh three ounces of flour without looking at the scales.

ESTIMATION   PROOF

How much does an egg weigh?

ESTIMATION   PROOF

Estimating sizes, distances, and weights
is a complex memory exercise.

Try to answer the above questions as correctly
as possible, then test your answers.

Concentration Exercises

126 TWISTED!

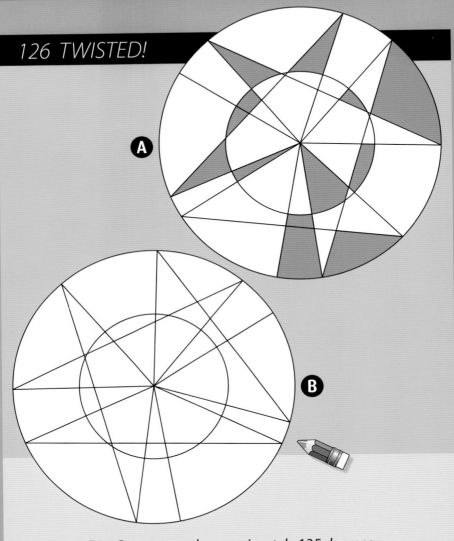

*Disc B was turned approximately 135 degrees
to the left compared to disc A.
Color disc B just like disc A.*

Concentration Exercises

127 LEARN CHINESE

1	2	3	4	5
一	二	三	四	五

6	7	8	9	10
六	七	八	九	十

52 → 五十二   　十三 → 13

60 →　　　　　　　十一 →

76 →　　　　　　　三十八 →

99 →　　　　　　　五十八 →

Here are the Chinese numbers 1 through 10 and the numbers 52 and 13. Based on this information, translate the numbers above to Chinese and the Chinese numbers to western (Arabic) numerals.

Concentration Exercises

128 3D-EXERCISE

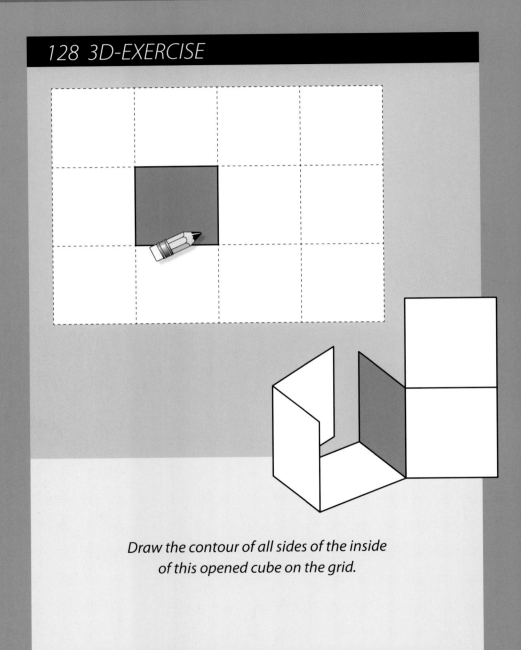

*Draw the contour of all sides of the inside
of this opened cube on the grid.*

Concentration Exercises

129 ILLUSION

What else do you see other than a bunch of twos and fives?

Concentration Exercises

130 WITH ONE LINE

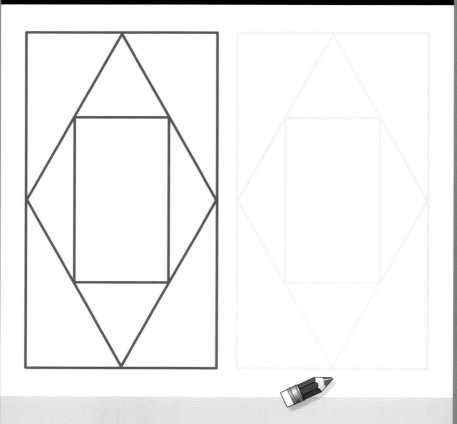

Try to draw this shape with one continuous line without lifting your pencil off the page and without any overlapping. There may be more than one solution.

Concentration Exercises

131 WITH ONE LINE

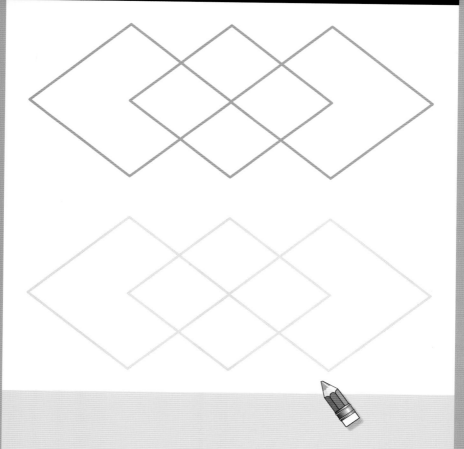

Try to draw this shape with one continuous line without lifting your pencil off the page and without any overlapping. There may be more than one solution.

Concentration Exercises

132 LATERAL THINKING

Connect the nine dots with one continuous line.
You can only change direction three times.

Concentration Exercises

133 SHIFT PUZZLE

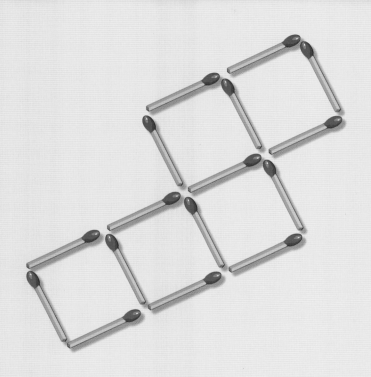

Move two matches to obtain four squares.

Concentration Exercises

134 LOCOMOTION

B

1.
Place the tip of your pen on the red dot A and then on the white dot B. Do this ten times in a row.

2.
Now place your pen on dot A and close your eyes. Try to indicate the center of dot B as accurately as possible with your eyes closed.

Concentration Exercises

135 FIND THE NINE DIFFERENCES

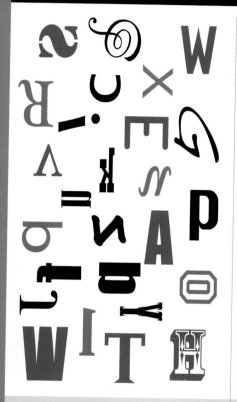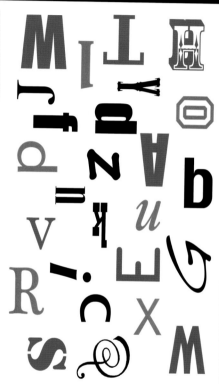

Find the nine differences in the right mirror image.

Concentration Exercises

136 A STAR IS BORN

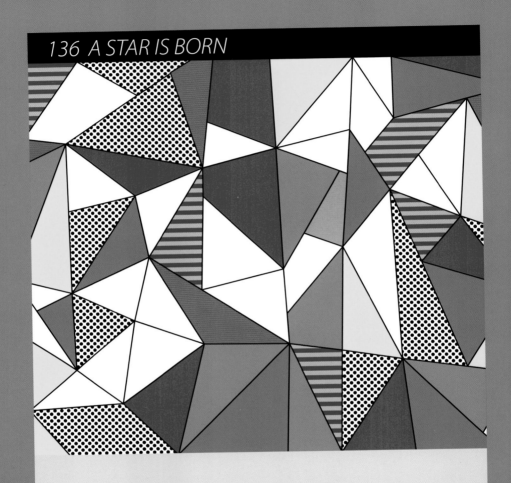

Where is this star ☆ *hidden in the drawing?*

Concentration Exercises

137 DIVIDED FAIRLY

Draw three straight lines from side to side
so that five sections are created, each one
containing the first five letters of the alphabet.

Concentration Exercises

138 WATCH OUT FOR THE ILLUSION

Which two tables (1–5) have identical tabletops?

Concentration Exercises

139 3–D EXERCISES

*Which profile (1–4) of the same house
is shown incorrectly?*

Concentration Exercises

140 WATCH OUT FOR THE ILLUSION

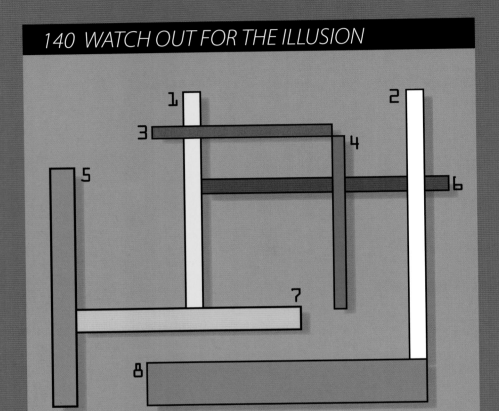

Which sets of lines are the same length,
thereby forming four pairs?

Concentration Exercises

141 COUNT(RY) MUSIC

1+2=

13-4=

7x8=

27/3=

¼ of 100=

136+24=

67-100=

12x4=

4444/11=

²/₆ of 30=

(-4)+16+(-12)=

38-(-24)-22=

(8x2)-(60/5)=

(12/4)-(-3x9)=

$$\frac{[(18/3)x(6/3)]+12}{(3x7)+3} =$$

With your favorite music blaring in your headphones,
try to solve the following calculations without a calculator.

Concentration Exercises

142 MAZE

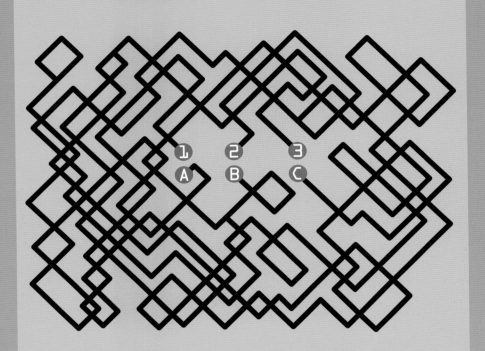

Spot the number that you can connect with a letter.
You can only go straight, so at an intersection you cannot
change directions. Only use your eyes don't use your finger.

Concentration Exercises

143 SPOTTING SQUARES

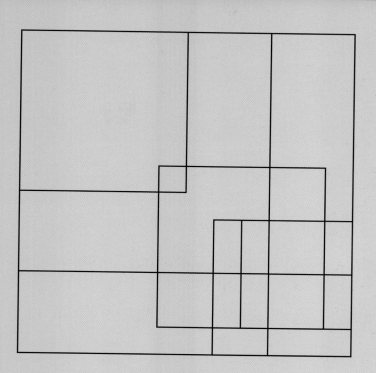

How many squares do you see?

Concentration Exercises

144 IDENTICAL PARTS

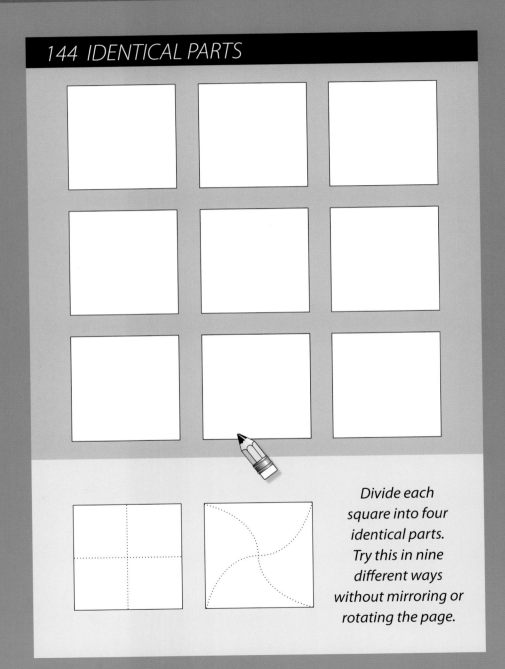

Divide each square into four identical parts. Try this in nine different ways without mirroring or rotating the page.

Concentration Exercises

145 SHIFT PUZZLE

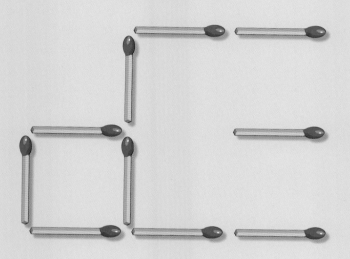

Move three matches to obtain two squares.

GRID PUZZLES

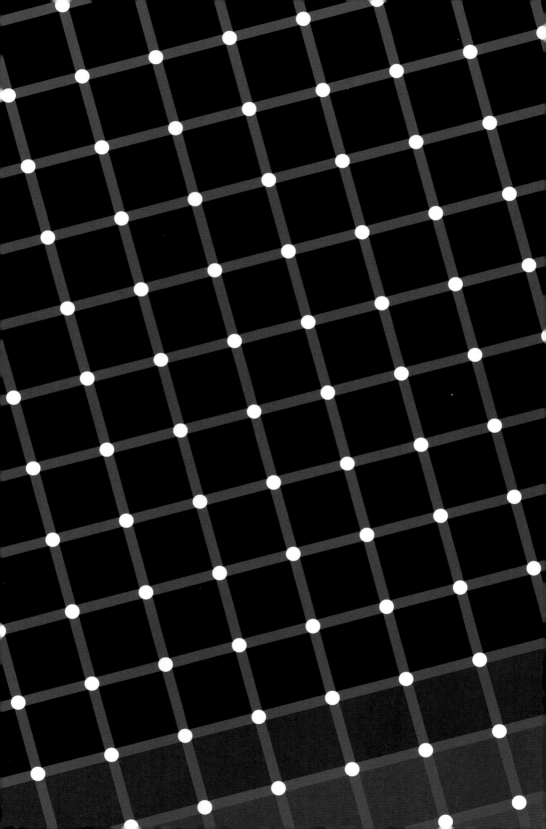

Relaxation in the game

Puzzles and riddles have always been around. Though we cannot prove it, we assume that early man viewed puzzles as a form of entertainment.

Don't cherish any illusions: Solving a grid puzzle requires practice and concentration. This is partially inherent, which provides a form of relaxation. A pleasant side effect is that the more often you play, the more proficient you become, and that gives more satisfaction once you discover the underlying strategy.

To solve a grid puzzle you'll have to use all your capacity for logical thought. It might not work the first time, but it will probably work the next time. Start with the easy ones and slowly proceed to the more difficult puzzles. After a while, you'll be mad about them.

Tip!
Some puzzles in this chapter are not easy. That's why identical, smaller versions are at the bottom of the page. Use them to test certain choices or strategies.

Grid Puzzles

DIRECTIONS PIXELFUN

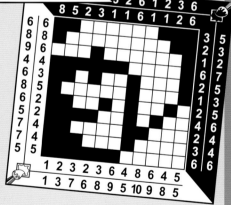

Total number of black squares

Total number of white squares

Largest group of adjoining white squares

Largest group of adjoining black squares

Color the correct squares black and discover the pixel image.

The numbers on the outer ring against the black or the white background indicate the total number of black or white squares on a column or a row. The numbers on the inner ring against the black or white background indicate the largest group of adjacent black or white squares to be found anywhere on that column or row.

For instance, if there is a six on the outer ring and a two on the inner ring against a white background, then there are six white blocks in that row, and the biggest group or groups consist of a maximum of two adjacent white blocks.

Grid Puzzles

146 PIXELFUN

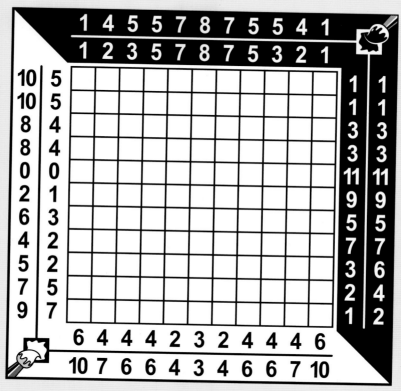

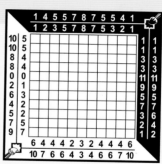

TEST ZONE

Grid Puzzles

147 PIXELFUN

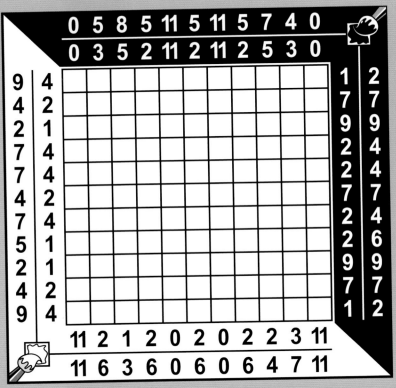

TEST ZONE

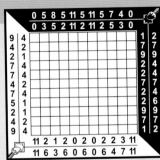

Grid Puzzles

148 PIXELFUN

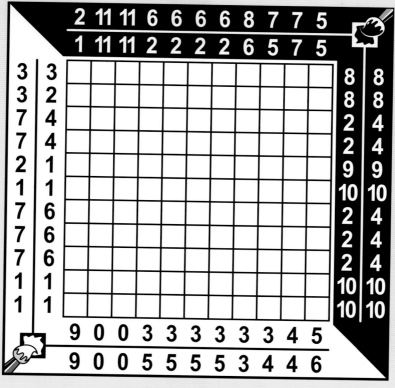

TEST ZONE

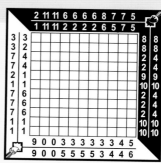

Grid Puzzles

149 PIXELFUN

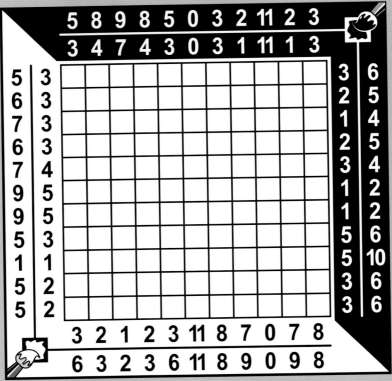

TEST ZONE

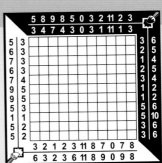

Grid Puzzles

150 PIXELFUN

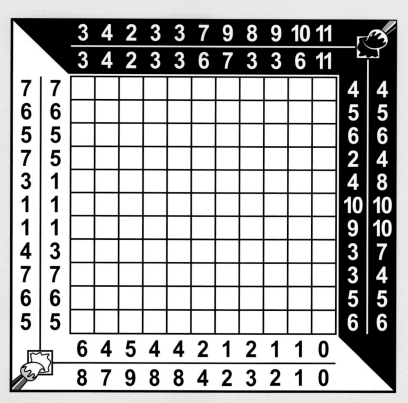

TEST ZONE

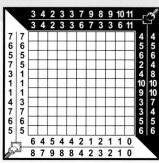

Grid Puzzles

151 PIXELFUN

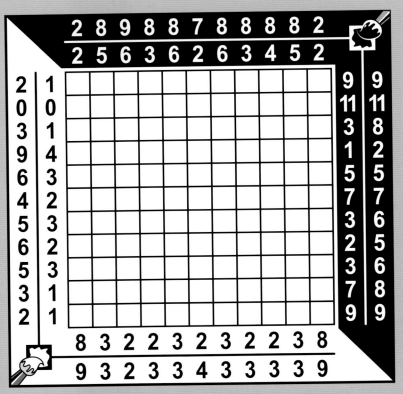

TEST ZONE

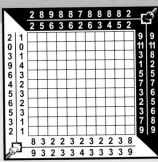

Grid Puzzles

152 PIXELFUN

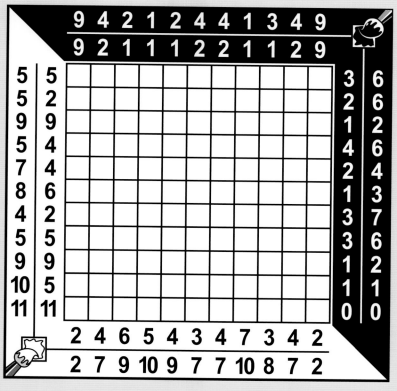

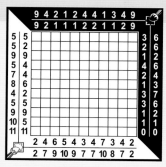

TEST ZONE

Grid Puzzles

153 PIXELFUN

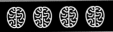

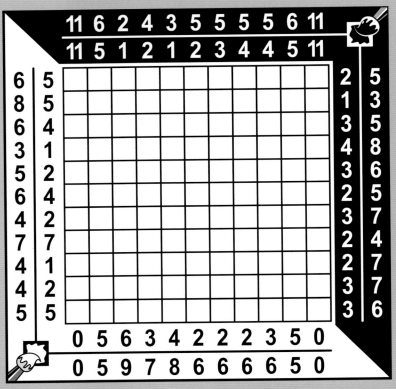

TEST ZONE

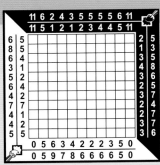

Grid Puzzles

154· PIXELFUN

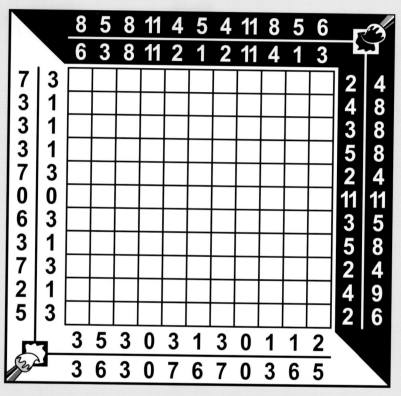

TEST ZONE

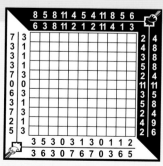

Grid Puzzles

155 PIXELFUN

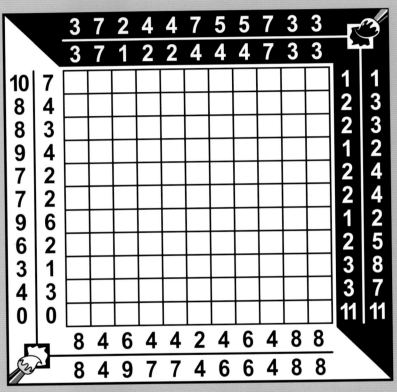

TEST ZONE

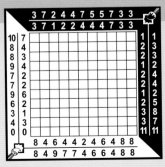

Grid Puzzles

156 PIXELFUN

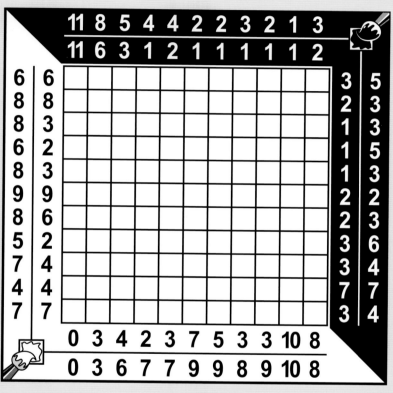

TEST ZONE

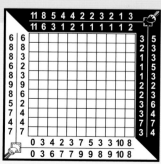

Grid Puzzles

DIRECTIONS CONTINUOUS LINE

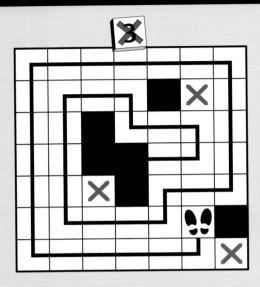

1. Start on a blank square of your choice and connect as many blank squares as possible with one single continuous line.

2. You can only connect squares along vertical and horizontal lines, not along diagonal lines. You must continue the connecting line up until the next obstacle, i.e. the rim of the box, a black square, or a square that has already been used.

3. You can change directions at any obstacle you meet.

4. Each square can only be used once. The number of blank squares that will be left unused is marked in the upper square.

5. There is more than one solution. We only show one solution.

Grid Puzzles

157 CONTINUOUS LINE

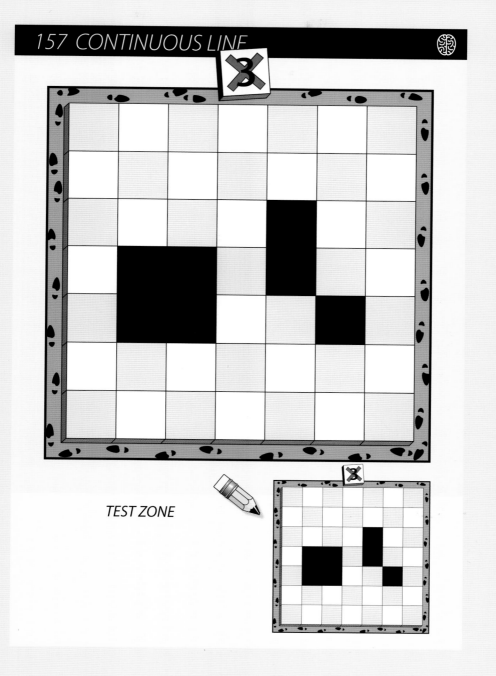

TEST ZONE

Grid Puzzles

158 CONTINUOUS LINE

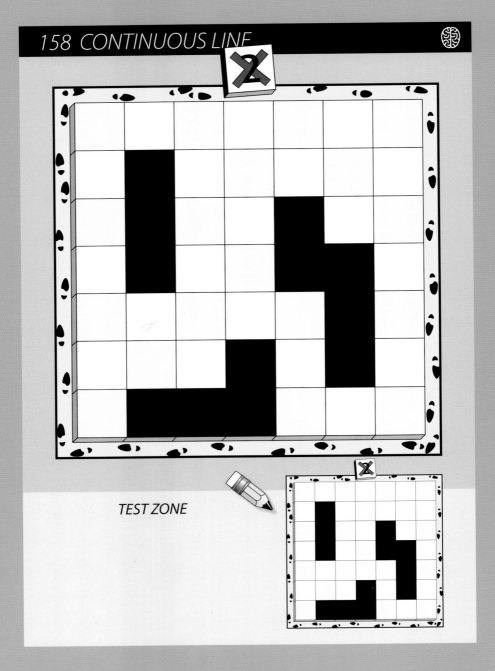

TEST ZONE

Grid Puzzles

159 CONTINUOUS LINE

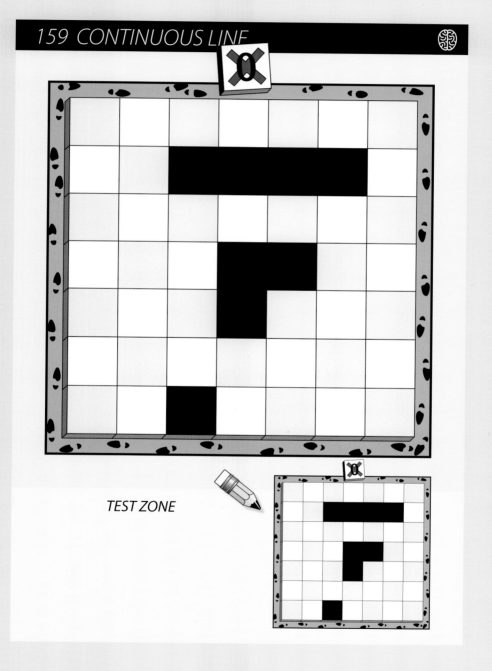

TEST ZONE

Grid Puzzles

160 CONTINUOUS LINE

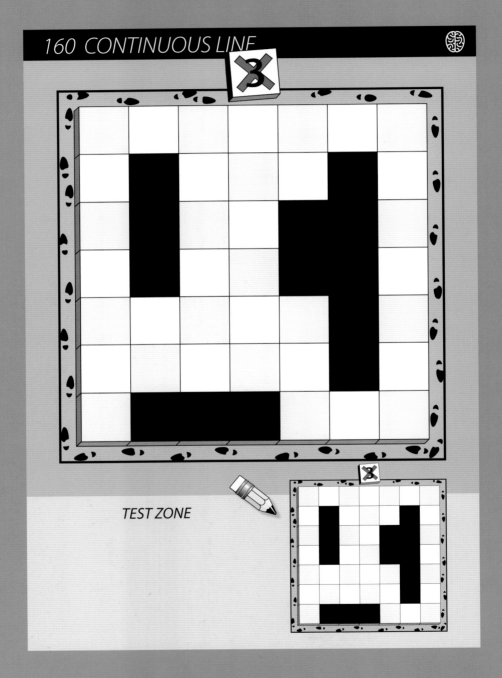

TEST ZONE

Grid Puzzles

161 CONTINUOUS LINE

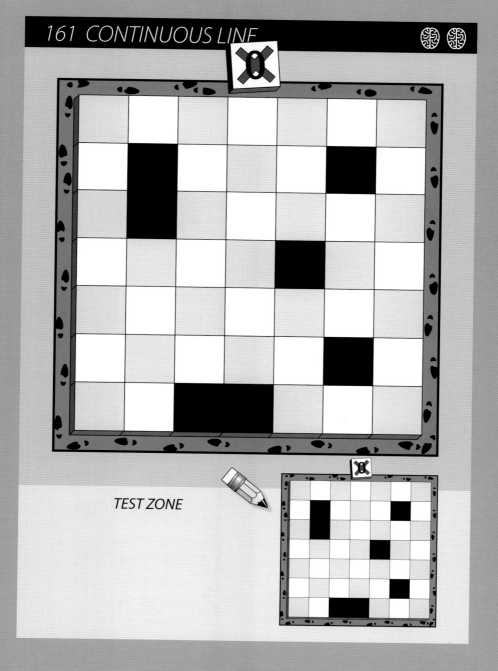

TEST ZONE

Grid Puzzles

162 CONTINUOUS LINE

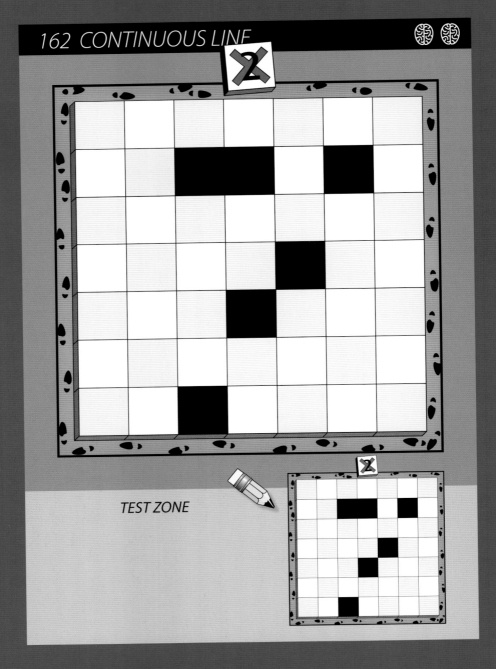

TEST ZONE

Grid Puzzles

163 CONTINUOUS LINE

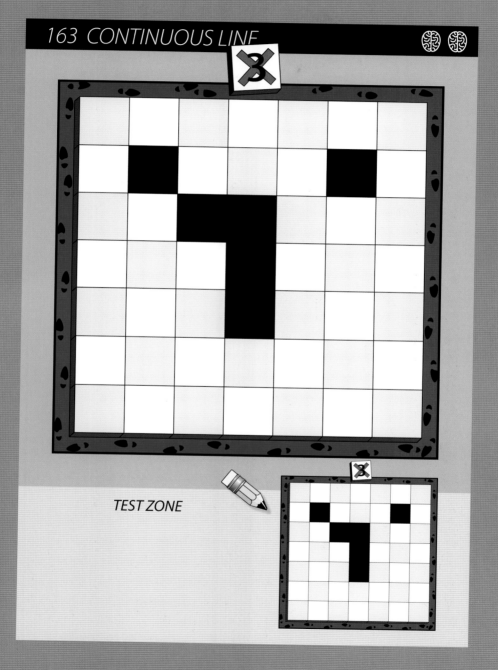

TEST ZONE

Grid Puzzles

164 CONTINUOUS LINE

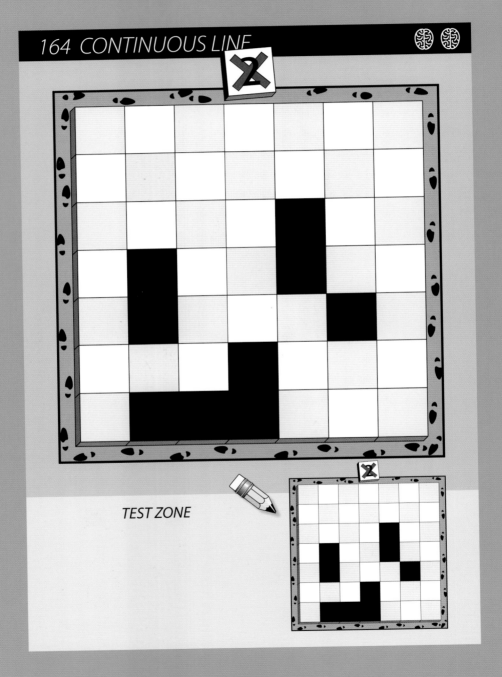

TEST ZONE

Grid Puzzles

165 CONTINUOUS LINE

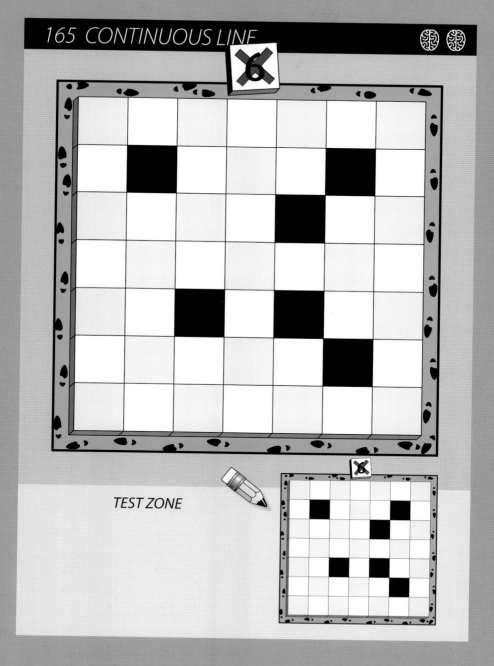

TEST ZONE

Grid Puzzles

166 CONTINUOUS LINE

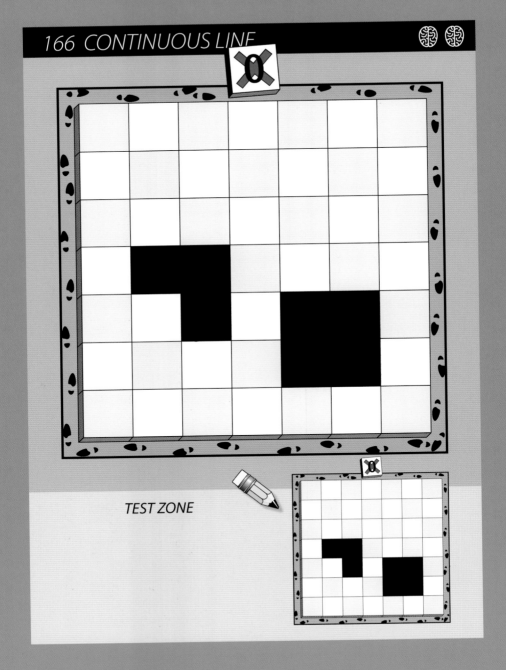

TEST ZONE

Grid Puzzles

167 CONTINUOUS LINE

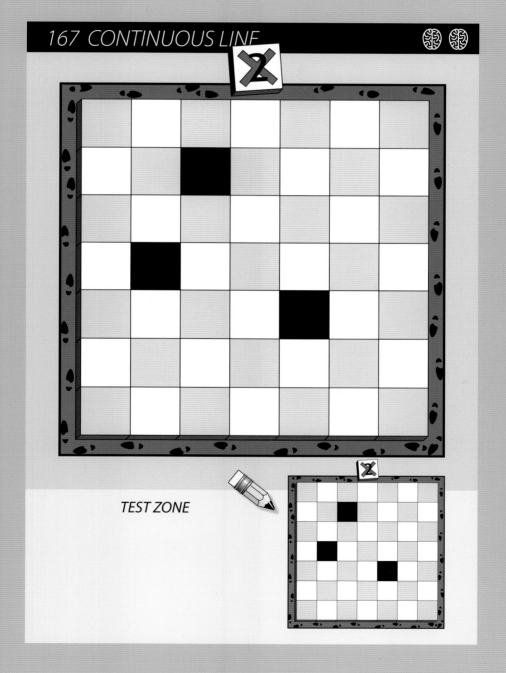

TEST ZONE

Grid Puzzles

DIRECTIONS ZOO

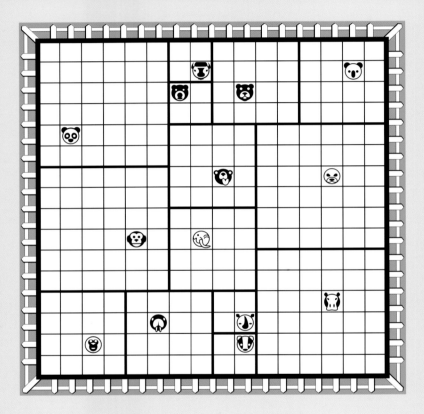

Cage the animals.

1. *Draw the lines that will completely divide up the grid in small squares with exactly one animal per square.*

2. *The squares should not overlap.*

Grid Puzzles

168 ZOO

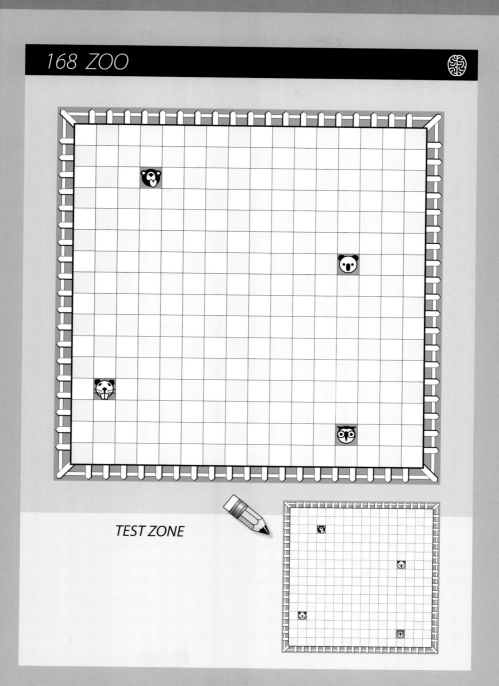

TEST ZONE

Grid Puzzles

169 ZOO

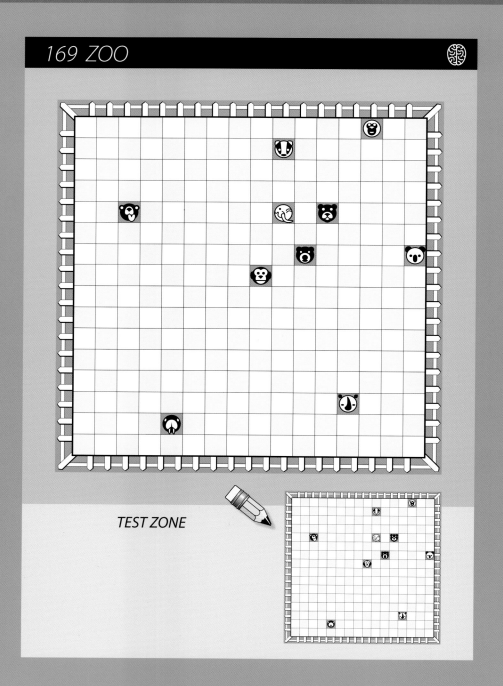

TEST ZONE

Grid Puzzles

170 ZOO

TEST ZONE

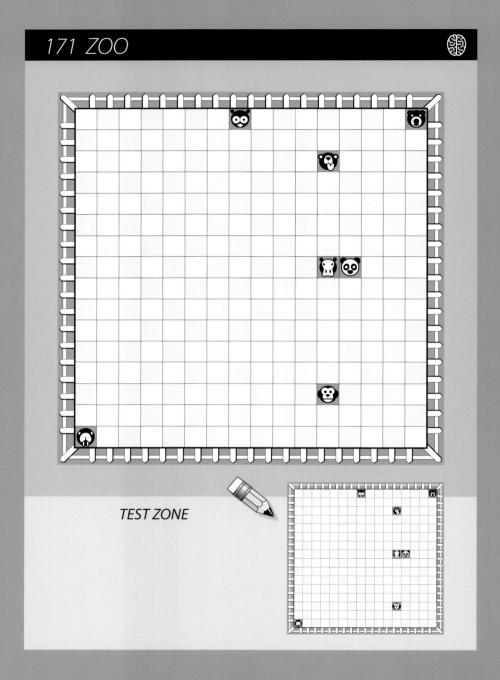

Grid Puzzles

171 ZOO

TEST ZONE

Grid Puzzles

172 ZOO

TEST ZONE

Grid Puzzles

173 ZOO

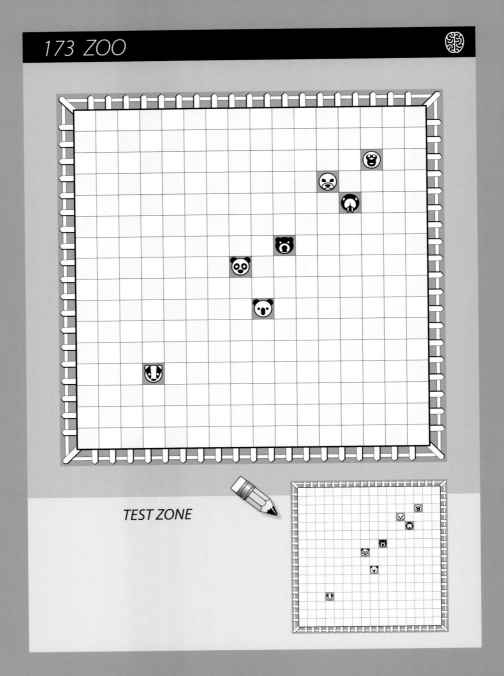

TEST ZONE

Grid Puzzles

174 ZOO

TEST ZONE

Grid Puzzles

175 ZOO

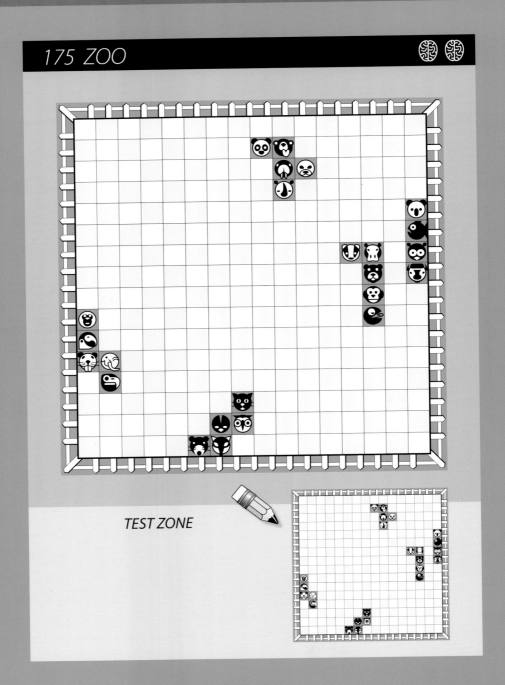

TEST ZONE

Grid Puzzles

176 ZOO

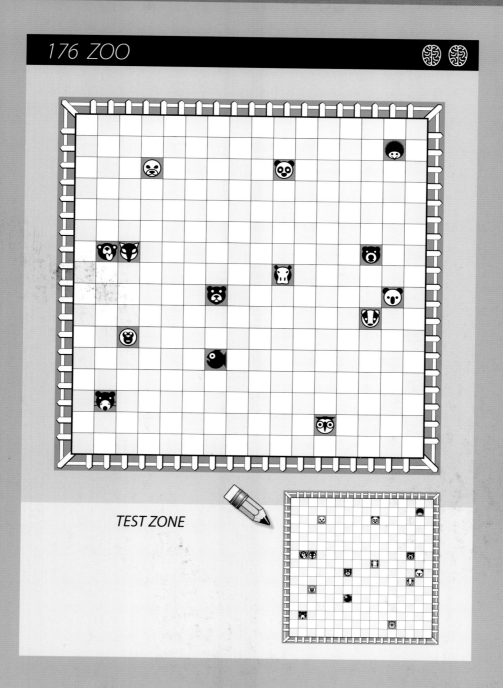

TEST ZONE

Grid Puzzles

177 ZOO

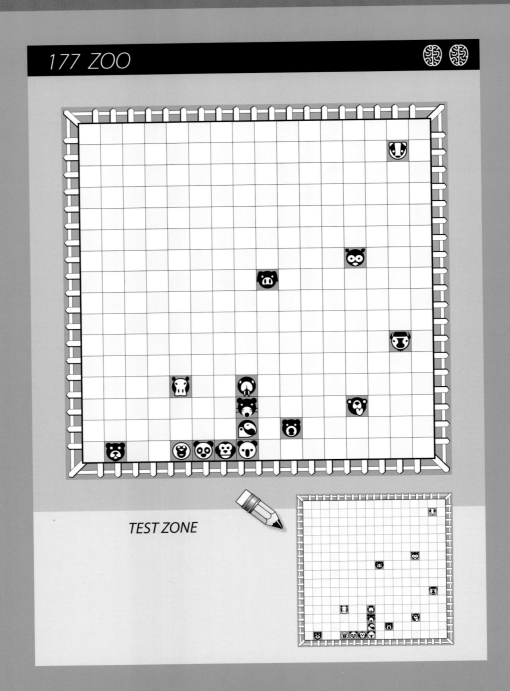

TEST ZONE

Grid Puzzles

178 ZOO.

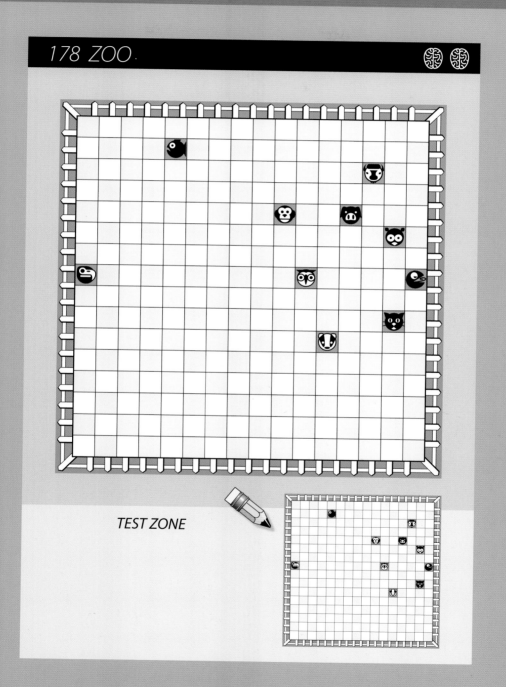

TEST ZONE

Grid Puzzles

DIRECTIONS FIND THE DIAMONDS

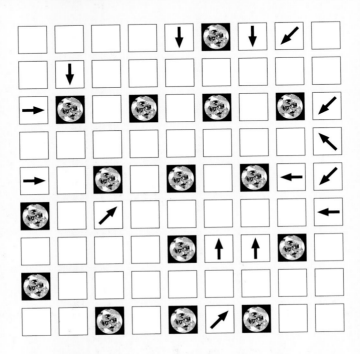

1. Knowing that every arrow points to another diamond and that no diamond can touch another vertically, horizontally, or diagonally, find the missing diamonds.

2. A diamond cannot be located on an arrow.

3. We show one diamond.

Grid Puzzles

179 FIND 15 DIAMONDS

Grid Puzzles

180 FIND 15 DIAMONDS

① ② ③ ④ ⑤ ⑥ ⑦ ⑧ ⑨ ⑩
⑪ ⑫ ⑬ ⑭ ⑮

Grid Puzzles

181 FIND 13 DIAMONDS

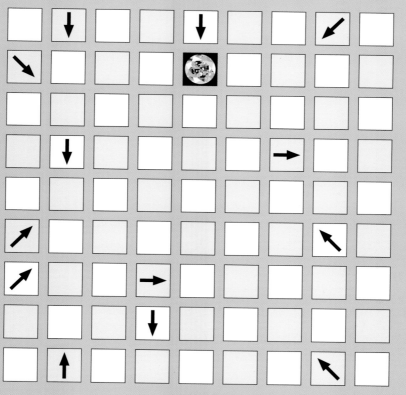

Grid Puzzles

182 FIND 15 DIAMONDS

① ② ③ ④ ⑤ ⑥ ⑦ ⑧ ⑨ ⑩

⑪ ⑫ ⑬ ⑭ ⑮

Grid Puzzles

183 FIND 15 DIAMONDS

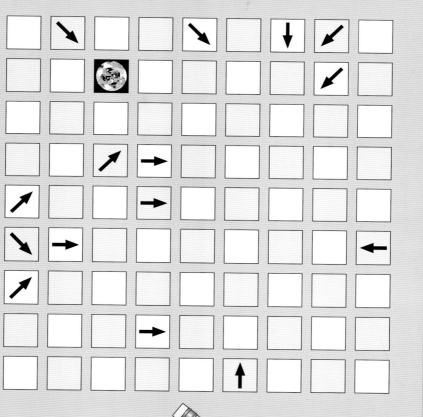

Grid Puzzles

184 FIND 15 DIAMONDS

① ② ③ ④ ⑤ ⑥ ⑦ ⑧ ⑨ ⑩
⑪ ⑫ ⑬ ⑭ ⑮

Grid Puzzles

185 FIND 16 DIAMONDS

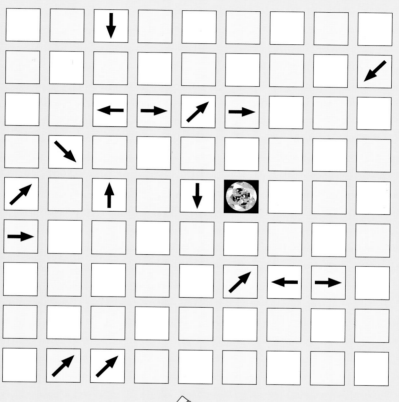

Grid Puzzles

186 FIND 16 DIAMONDS

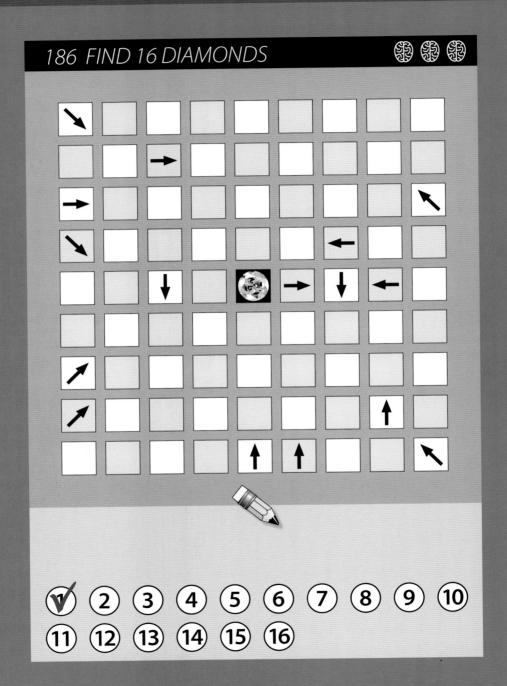

Grid Puzzles

187 FIND 15 DIAMONDS

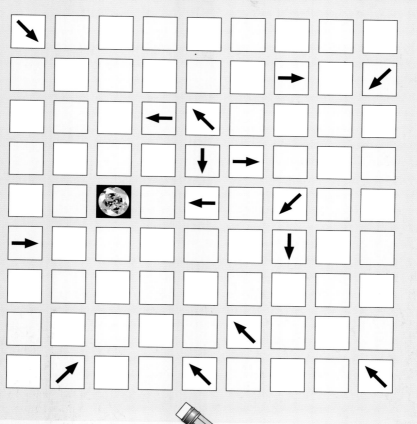

① ② ③ ④ ⑤ ⑥ ⑦ ⑧ ⑨ ⑩
⑪ ⑫ ⑬ ⑭ ⑮

BRAIN TWISTERS, MIND BENDERS,

Grid Puzzles

188 FIND 15 DIAMONDS

① ② ③ ④ ⑤ ⑥ ⑦ ⑧ ⑨ ⑩
⑪ ⑫ ⑬ ⑭ ⑮

Grid Puzzles

189 FIND 15 DIAMONDS

① ② ③ ④ ⑤ ⑥ ⑦ ⑧ ⑨ ⑩
⑪ ⑫ ⑬ ⑭ ⑮

Grid Puzzles

DIRECTIONS BINARY

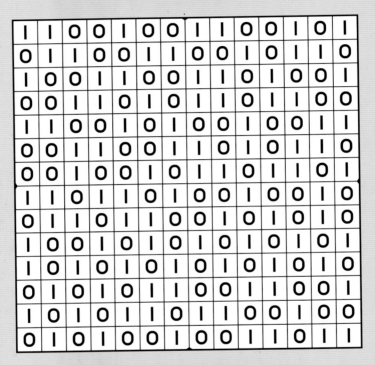

1. Complete the grid with zeros and ones until there are just as many zeros and ones in every row and every column.

2. No more than two of the same number can be next to or under each other.

3. Rows or columns with exactly the same content are not allowed.

Grid Puzzles

190 BINARY 12x12

	O							O			
O			I								
				O					O	O	
		O			I	I					O
O		I								O	
					I		O	O			
				I							
				I		I					O
						O		I	I		
		O			I			I			
				O			O				
O		O	O			I					I

Grid Puzzles

191 BINARY 12x12

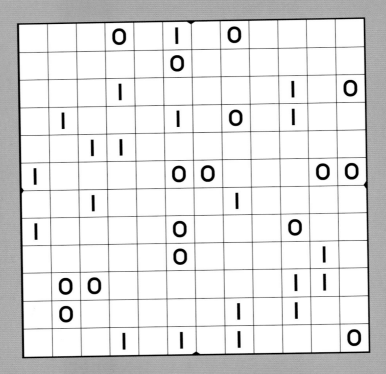

Grid Puzzles

192 BINARY 12x12

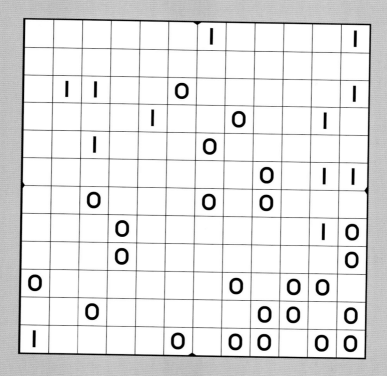

Grid Puzzles

193 BINARY 12x12

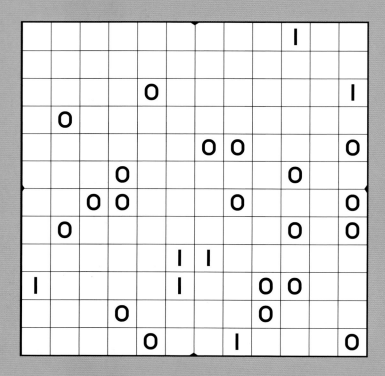

Grid Puzzles

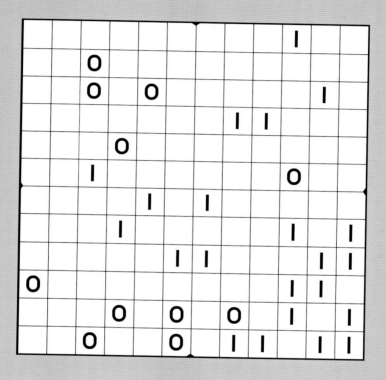

Grid Puzzles

195 BINARY 14x14

						O			O		I		
O										O			
				O			O			O	O		
				O		I	I				O		
		I	I										
I						I				I		I	
I		I						O					
					O					O		O	
		O	O			I		I	I				
											O		
O			I			I		O					
				I		I		O	O		I	I	
			O							I	I		
		O	O			I		O		I	I		I

246

Grid Puzzles

196 BINARY 14x14

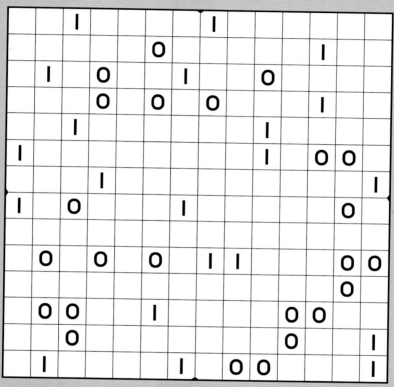

Grid Puzzles

197 BINARY 14x14

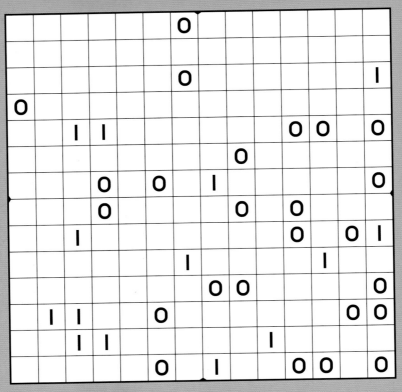

Grid Puzzles

198 BINARY 14x14

			O										
		O							I				O
				O			I				I	I	
O		I	I			O			I				
	I		I						I		O		
		O					O				O		O
			I						O				
							I					O	
I		I	I										I
									O		I	I	
	O		O							O			
				O			I		I				
						O	I		I			O	O

Grid Puzzles

199 BINARY 16x16

			O		O			I							
		I													
					O		O		I		O			O	
	O			I	I			O							
		I		I	I					O		I			
											O		O		
		I								I			O		
			I	I		I	I		O		O		O		
		I		I		I	I								I
O											I				
	O			O	O							I			
					I						O		O		
O		O	O				I				O				I
	I				I										I
		O	O			O		I	I						
O			O			O			I				I	I	

Grid Puzzles

200 BINARY 16x16

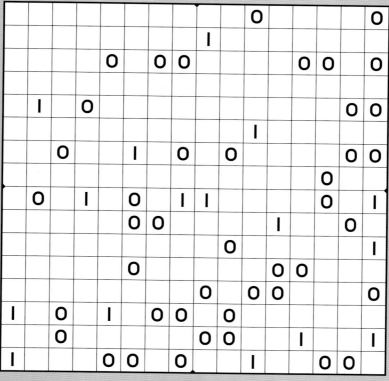

Grid Puzzles

201 BINARY 16x16

Grid Puzzles

202 BINARY 16x16

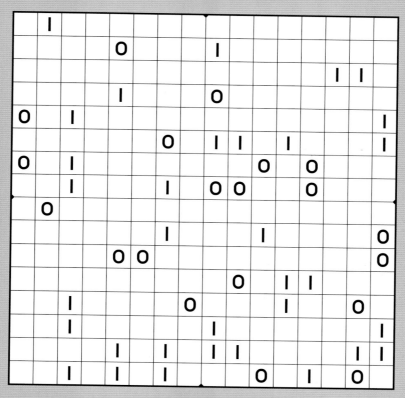

203 BINARY 18x18

Grid Puzzles

204 BINARY 18x18

Grid Puzzles

205 BINARY 18x18

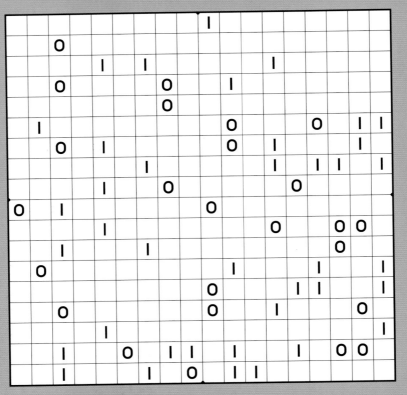

Grid Puzzles

206 BINARY 18x18

				O					I	I							
I	I			I		O		I		I				I	O		
							O			I		I		I		I	I
	O	O			O		O	O									
											I						
			I		O		O					O	O				
					I	I											
		O								I			I	I			
						O									I	I	
O									O		I	I					O
	I			I											I		
		O		I				I		I							
											O		O	O		O	O
				I													
	I				O		I	I							O		
	I		O								O	O					I
I																I	O

Solutions

1   Symbol 4
      Join the beginning and the end of the strip to see the rest.

2   6F
      Points are scored on intersections where the number
      corresponds with the letter's place in the alphabet.

3   L
      We see the letters C to K alternating left and right.
      The next letter is L.

4   47 blocks
      22 on the bottom and 25 for the superstructure.

5   1
      The number in the circle corresponds to the
      number of different types of candy in the set.

6   Six of hearts
      The same cards are used in every column.

7   6
      The next domino always has one dot more.

8   134

9   4

10   Train 4
      On all the other trains the same elements change color.

11   Insulation 4
      $A=B+C=2C+D$. So $B=C+D$.
      If we convert everything to C and D then the
      value for insulation A is as follows:
      $1=2A+C+D=2(2C+D)+C+C=5C+2D$.
      By analogy we find that
      $2=4C+2D$, $3=3C+2D$, $4=5C+3D$, $5=5C+2D$, and $6=5C+2D$.

Solutions

12   Paint 3
     Paint 1 was used twice to color in 50 percent of a surface.
     Paint 2 was used once to color in 50 percent and
     once to color in less than 50 percent.
     Paint 3 was used once to color in 50 percent and
     once to color in more than 50 percent.

13   They will all be together on line 19.

14   11
     Replace the compass rose with a clock on military time.
     The red tip of the compass needle points to the
     hour in the afternoon from 13 to 24 midnight and
     corresponds with the day temperature.
     The gray tip points to the hour in the morning from
     1 to 12 noon and is the night temperature.

15   The race started at 09:58:29 and ended at 10:01:00.
     Keep in mind:
     A. The number design
     B. There are 60 seconds in one minute, 60 minutes in one hour,
     C. You have to add 2 minutes and 31 seconds.
     1. Possible seconds
        The start seconds can only be 00–09 and 20–29. (see A and B)
        The end seconds can only be 00 and 20–29. (see A and B)
        31 seconds+start seconds>29, so the end seconds must be 00.
        only 29+31=60 seconds or 1 minute and 00 seconds.
     2. Possible hours
        Start hours can only be 2, 3, 4, 5, 6, 8, 9
        End hours can only be 10 or 11.
        With 2 minutes and 31 seconds you can
        make the hours switch from 9 to 10.
     3. Possible minutes
        The start minutes can only be 58 (see A and B)
        The end minutes can only be
        58+2+1=61 seconds or 1 hour and 1 minute

16   With group 2 and 3

Solutions

17   *auteb*
All words are underlined in which the first and last letters are consecutive in the alphabet.

18   *Butterfly C*
A, B, and D have the three family colors.

19   *D*
The GPS counts one turn further at every roundabout.
If it counts more than the number of turns, it continues to ride on the roundabout until it comes to the correct number of turns.
Arriving at the first roundabout the GPS tells the driver to take the first turn (street).
Arriving at the next (second) roundabout the GPS tells the driver to take the second turn (street).
Arriving at the next (third) roundabout the GPS tells the driver to take the third turn (street).
Its route looks like this (roundabout/turn):
7/1-8/2-9/3-6/4-5/5-3/6-6/7-2/8-4/9-1/10.

20   *Rose 3*
All the other roses have thorns.

21   *Pumpkin 4*
All the other pumpkins have two eyes with just as many angles.
The right eye of pumpkin 4 has 6 angles; the left eye only has 5.

22   *Point 10*

23   *Cheese cube 3*
The mouse chooses the chunks of cheese with the fewest number of holes.
It already collected the cubes with 0, 1, 2, and 3 holes.
The cube with four holes is the next one.

24   *Zone 6*
Starting in the center of the cobweb, there is a yellow zone every three zones.

Solutions

25   *Group 2*
The sun is always behind the clouds, never in front of a cloud.

26   *Set 3*
All the other sets consist of only three different colors.

27   *HEK is the only three-letter word that does*
not appear in the name Shakespeare.

28   *Five people*
Four oarsmen and a coxswain.
You can discover this because there is not an extra
oar visible at the front of the top image.

29   *You encounter five chalets via TUSW.*
The other routes are ABCDE: four chalets
and KLMNOP: three chalets.

30   *On hill D*
If you count the contour lines you'll discover that all the
other windmills are located on the seventh contour line.

31   *Spray can 5*
It bears the three symbols with the highest value.

32   *Color 5*
The beads are split in groups of three.
The color of the middle bead is a combination
of the outer two colors.

33   *Direction B, clockwise*

34   *5 percent*
The percentage equals the number of fruits
shown on the container above.

35   *Two barrels*
The number is always reversed and the last number is dropped.

Solutions

36 Angle 5
The left front and back tower have to switch to be correct.
Also the right front and back tower have to switch to be correct.

37 Kebab 5
All other kebabs start and end with a sausage.

38 F=blue and H=yellow
This work of art consists of 6 identical square areas
that were each turned forty-five degrees.

39 Segments 2 and 3
In the first and third row, from left to right, a
black segment in added clockwise.
In the second row, from left to right, a black
segment in added counterclockwise.
When two black segments overlap they disappear.

40 Color 5
Every fourth triangle is red.

41 Brain cell 3

42 20
The number of items equals the value
of the first letter of the map.

43 Square 9
Orange comes from above, blue from the right,
purple from the left, and yellow from below.

44 6
Each number is equal to the total number
of cubes on the X, Y, and Z axes.

45 Ball 3
The ball always turns ninety degrees and changes
color. The valve is on the top of the ball.

Solutions

46   Pen 2
On all the other pens the clip of the cap has the color
of the largest number of identically colored dots and
the cap is the same color as the remaining dot.

47   54321
Compare the number of flags with the
number of letters to find the solution.

48   Card B, five of spades
The corners of A are not correct. C is already in the hand.
D contains nine diamonds instead of seven.

49   Diagram 3
The orange block always moves one spot to the
right and alternates between low and high.

50   7L1IC4
The vowels and even numbers are red in all the other numbers.

51   6565656
The parking lot was filled from top to bottom and from left to
right. One additional car is added each time the color changes
and the cars alternate being parked nose inward and outward.

52   Ice cube 5

53   Spinning movement 3
All other spinning movements consist of three different
patterns. Number 3 has are only two different patterns.

54   Ghost 3
The eyes are shows in reverse on the
tombstones of all the other ghosts.

55   Chick D

56   Lip imprint 8

Solutions

On the other imprints pink is only used for the upper lip and black for the lower lip. This is reversed in imprint 8.

57 *843*
The yacht always sails to the next island that is alternately the farthest east or farthest west.

58 *Patch 4*
The space between the patches forms the word TEST.

59 *2*
The detective only uses lines that branch off.

60 *Number 10*
Two inspection robots are always looking at each other.

61 *X and E*
All the letters are arranged alphabetically and diagonally, from the lower left corner to the upper right corner.

62 *Five types*
The three on the top of each pile, the bottom one, and the second cap from the top starting from the pile in the middle.

63 *No solution*

64 *No solution*

65 *No solution*

66 *No solution*

67 *Mnemonic device:*
You can associate the words per group with one color.

68 *No solution*

69 *No solution*

Solutions

70   Mnemonic device:
Connect the dots according to the colors of the rainbow.

71   No solution

72   No solution

73   No solution

74   During pregnancy, *babies* make up to three thousand brain cells per *minute* during certain periods. Once fully grown, the brain is composed of about a hundred *million* neurons. Between each cell and hundreds of neighboring *neurons* up to ten thousand *synoptic* connections can be made as we learn something new. These first weak connections become *weaker* upon repetition or when intense emotions influence the learning process.

75

	Ann	Elsie	Lucie	Bea	Bart	John	Tom	Peter
Ann	♡	♡	♡	✗	♡	♡	♡	♥
Elsie	♡	♡	♥	♡	♡	♡	♡	✗
Lucie	♡	♥	♡	♡	✗	♡	✗	♡
Bea	✗	♡	♡	♡	♡	♡	♡	♥
Bart	♡	♡	✗	♡	♡	♥	✗	♡
John	♡	♡	♡	♡	♥	♡	♡	♡
Tom	♡	♡	✗	♥	✗	♡	♡	♡
Peter	♥	✗	♡	♡	♡	♡	♡	♡

76   No solution

Solutions

77   No solution

78   No solution

79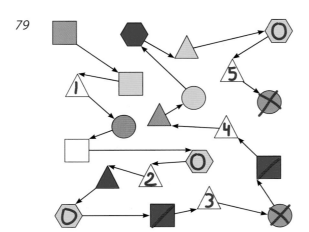

80   $6=2+2+2=(3x3)-3=\sqrt{(4x4)}+\sqrt{4}=(5/5)+5=$
     $6+6-6=7-(7/7)=8-\sqrt{(\sqrt{(8+8)})}=\sqrt{9}x\sqrt{9}-\sqrt{9}$

81   P=number of pigs   C=number of chickens
     P+C=72   4P+2C=200
     or 2P+C=100   or P=100-72(P+C)
     So there are a total of 28 pigs and 72-28=44 chickens.

82   The solution is:
     　　　2
     　　7　9
     　　3　　　4
     　8　1　6　5

83   6
     The difference in the scores forms the following series:
     -2, +4, -6, +8.
     Alternating per column you also see the following two series:
     7-14-21-28 and 9-18-27-36.

Solutions

84   3 spots
Read every row as follows:
13+23=36, 4+1=5, 2+1=3, 6+5=11, 31+33=64.

85   14
The egg white is worth 10 and the yolk is worth 1.

86   12
The number after the decimal point is half of the number before the decimal point and the decimal point is dropped. 24.12.

87   He needs a water depth of minimum
6 feet (draught)+3 feet (free depth)=9 feet. As of 8:00 A.M. the water increases by one foot per hour starting at five feet. In other words after four hours he will have enough water depth. 8:00 A.M.+four hours=12:00. Because the table was drawn up in wintertime and it is daylight savings time on August 18 he must add one hour. Therefore he can only enter the marina after 1:00 P.M..

88   2
From left to right the sum of two numbers always equals ten.

89   1100
After 0111 (7) comes 1000 (8).
If you add 0100 (4) to it you get 1100 (12).

90   Read as 8/2=4, 9/3=3, 6/1=6, 2/2=1.

91   This was a trick question. Any figure of three digits will do.

92   Nuclei 2 and 3 are equal because A=B
therefore blue=2x red+1x yellow.
Nucleus 2=2x(2x red+1x yellow)+purple.
Nucleus 3=2x(2x red+1x yellow)+purple.

Solutions

93   B=boy, H=hamburger, M=minutes
Rule of three
   1.5B eats 1.5H
   1B eats 1.5/1.5=1H
   6B eats 1Hx6=6 hamburgers in 1.5 minutes.
The same but now with the time
   In 1.5M 6 hamburgers are eaten.
   In 1M 6/1.5=4 hamburgers
   In 6M 4x6=24 hamburgers

94   30 points
The outer zone is worth 1 point, then 2, 3, etc.
The center is worth 10 points. 5+8+8+9=30.

95   After twelve hours
Clock A runs twenty minutes behind per hour so
it will indicate a full hour after three hours.
Clock B runs fifteen minutes. behind per hour so it will
indicate a full hour after four hours. Since twelve is
divisible by four and three, it will take twelve hours before
both clocks indicate a full hour at the same time.
On clock A it will be 4:00 and on clock B it will be 5:00.

96   It's impossible to weigh 5 ounces correctly using only
one weighing. All the other weights are possible.
For 1, 3, 4, 10, 11, 13, and 14 ounces you can
use the weights on one side of the scale.
You weight the other ounces as follows:
2 oz+1=3, **6 oz**+1+3=10, **7 oz**+3=10,
8 oz+3=10+1, **9 oz**+1=10, **12 oz**+1=3+10.

97   2
Starting at the top, the number of liters
of water is doubled every time.

98   890
The result is 128+79894-35604=44418.

Solutions

99   (P)eter+(L)inda+(E)lla=120
6E+3E+E=120,
so E=12
Peter is 6x12=72 and Linda is 3x12=36

100   183 - 294 - 675

101   On 18
The tokens are placed on numbers whose
digits are equal when subtracted.
A: 6-2(26)=4=5-1(15).
B: 8=9-1(19).
C: 3-1(13)=2-0(20)=2=4-2(24).
D: 7=8-1(18).

102   B
Piggy bank B always receives 1/3 of 24.
All the other piggy banks receive less.
A: 12, C: 15, D: 6, E: 12, and F: 3.

103   6462
The sum of the digits of all the other binders is 14.

104   9
The sum of the numbers is written under every cube.

105   21
The sum of the numbers making up the hours, minutes, and
seconds equals the number on the rider's back, and all hours,
minutes, and seconds start with 0, 1, and 2 respectively.

106   1   fish - amphibians - reptiles - mammals
        2   1 - 4 - 8
        3   speed of light

107   1   SIXTEEN - SEVENTY
        2   THUNDER - TSUNAMI
        3   TEACHER - DENTIST

Solutions

108	False	3 / it is only 1.4 pints
		4 / it is 3,000

109	False	3 / already after nine seconds
		4 / from 1 to more than 268 miles/hr

110 About, ballot, fake, derate, faster, fail, grateful, hells, indemnity, jars, knees, laser, modem, nearer, oven, plunged, queer, race, table, token, units, real, whits, y-axis, packed, heroes.

111 1   ellipse=πab
   2   parallelogram=bh
   3   circle=πr^2
   4   triangle=$\frac{1}{2}bh$
   5   trapezium=$\frac{1}{2}(a+b)h$

112	1	7:30 A.M.	Melatonin production stops
	2	10:00 A.M.	Highest alertness
	3	3:30 P.M.	Fastest reaction time
	4	6:30 P.M.	Highest blood pressure
	5	10:30 P.M.	Bowels are restrained
	6	2:00 A.M.	Deepest sleep

113 Anabolic steroid, narcotics, beta blocker, hormones, insulin

114 1   Ice Age - Blue Sky Studios
      Shark Tale - DreamWorks
      Flintstones - Hanna-Barbera
   2   bianco (nero) - Italian
      blanco (negro) - Spanish
      branco (preto) - Portuguese
   3   Andy Warhol - pop art
      Pablo Picasso - cubism
      Constant Permeke - expressionism
   4   Burundi - Tanzania
      Burkina Faso - Mali
      South Africa - Namibia

Solutions

	5	tofu - soy milk
		quorn - mold
		seitan - wheat flour
115	1	MINIMUM - MAXIMUM
	2	BLOCKS - PUZZLES
	3	SKIMMER - BLENDER
116	1a	2b 3b 4c 5c
117	1	Apple - Eve = Adam
	2	Crime - hammer = clue (sounds like clew)
	3	Hand - boss = handkerchief
	4	Ocean - sport = beach ball
	5	Fruit - talking= peaching
	6	Tennis - Religion = service
	7	Food - Ping-Pong = table
	8	Brain - hard disk = memory
	9	Butterfly - tank = caterpillar
	10	Tomato - balls= soup
	11	Film - board = director
	12	Sun - smart = bright
	13	Tropical - coniferous tree = pineapple
	15	Coffee - computer = Java
118	1	Thundercloud: TH under Cloud
	2	Red-eyes
	3	Shaker: shake + r
	4	Halfway: half of way
	5	Father: fat HER
119	1	cut excesses
	2	intercourse
	3	motivation
	4	preschool
	5	conflict
	6	comparing answers

Solutions

120   *New Year's Day (Jan 1)*
Martin Luther King Day (Jan 18)
Super Bowl Sunday (Feb 7)
Valentine's Day (Feb 14)
International Women's Day (March 8)
St. Patrick's Day (Mar 17)
Easter Day (Apr 22)
Mother's Day (May 9)
Memorial Day (May 31)
World Environment Day (June 5)
Father's Day (June 20)
Labor Day (Sep 6)
Columbus Day (Oct 11)
World Food Day (Oct 16)
Halloween (Oct 31)
Winter Time (End of DTS) (Nov 7)
World AIDS Day (Dec 1)
Christmas (Dec 25)

121   *1*   *SQUARE - PYRAMID*
      2   *LINCOLN - MERCURY*
      3   *BROTHER - HUSBAND*

122   *28 consonants*
blue-yellow-gray-black-white-green-red-purple-orange

123   *First $((2^3)/2)/4=1$*
and then $1/(1/2)=2$

124

125   *No solution*

Solutions

126 *No solution*

127
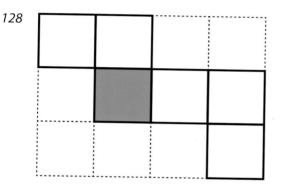
60 = 六十   十一 = 11
76 = 七十六   三十八 = 38
99 = 九十九   五十八 = 58

128

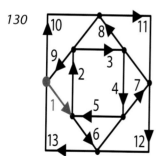

129 *All twos form an equilateral triangle.*

130

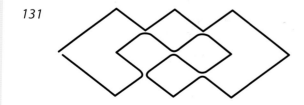

131

Solutions

132

133

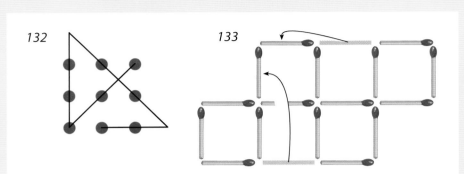

134 *No solution*

135

136

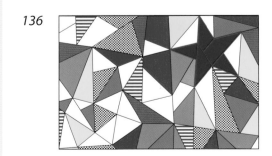

Solutions

137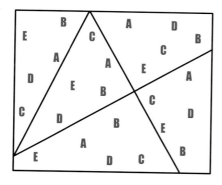

138 Tables 1 and 3

139 Profile 4
The drainpipe on the left should be in front of the planter.

140 Pairs 43, 56, 17, and 28

141 1+2=3
13-4=9
7x8=56
27/3=9
¼ of 100=25
136+24=160
67-100=-33
12x4=48

4444/11=404
²/₆ of 30=10
(-4)+16+(-12)=0
38-(-24)-22=40
(8x2)-(60/5)=4
(12/4)-(-3x9)=30
$\frac{(18/3)x(6/3)+12}{(3x7)+3}=1$

142 1C

Solutions

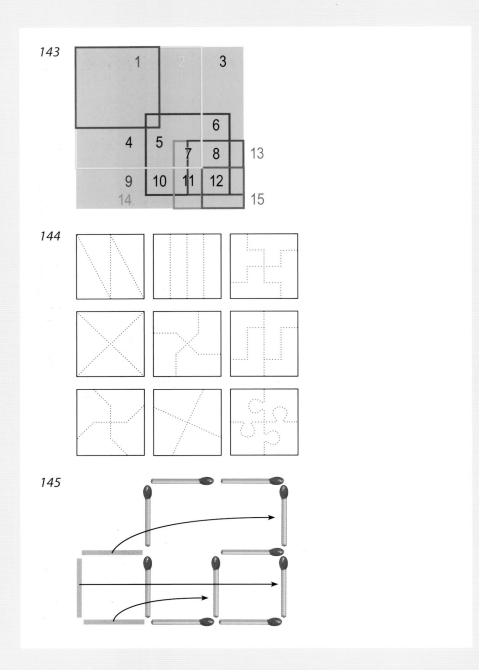

143

144

145

Solutions

Pixelfun

146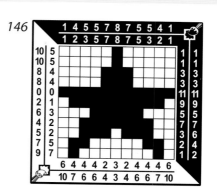

147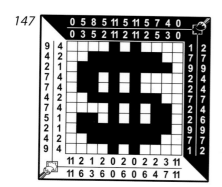

148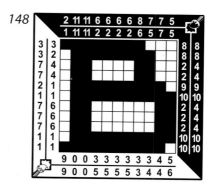

149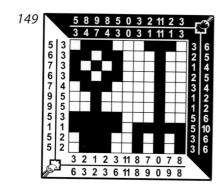

150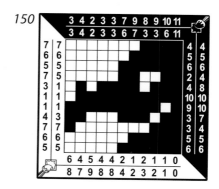

Solutions

151
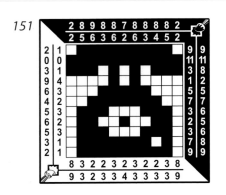

152
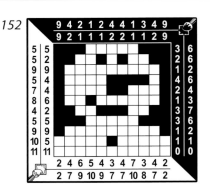

153
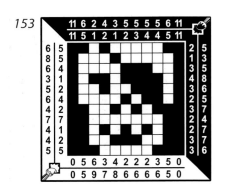

154
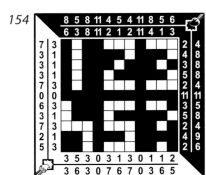

155
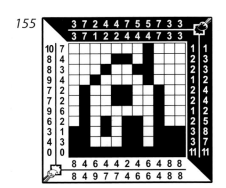

156
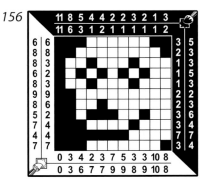

Solutions

Continuous line

157
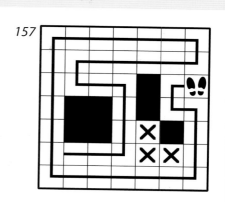

158
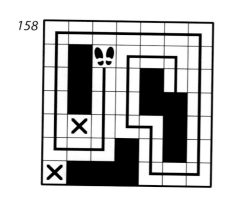

159
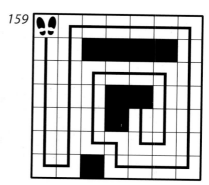

160
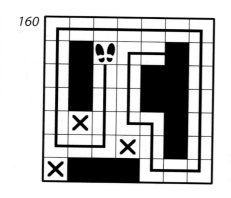

161
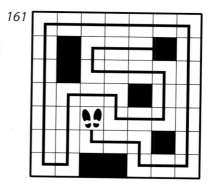

Solutions

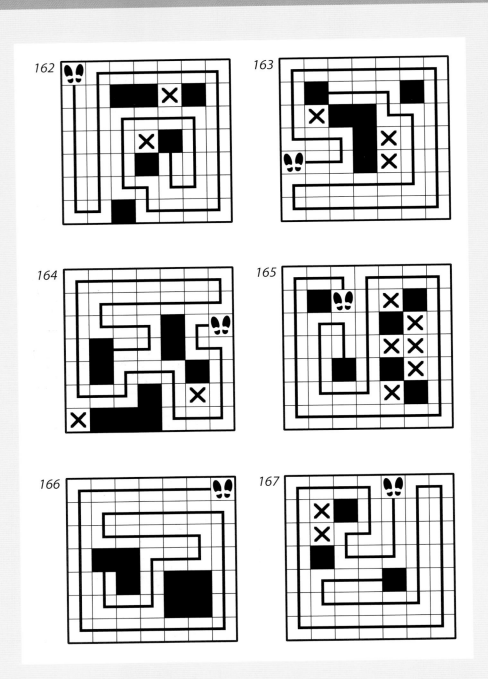

162

163

164

165

166

167

Solutions

Zoo

168
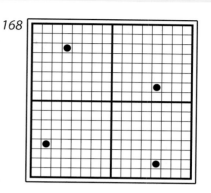

169
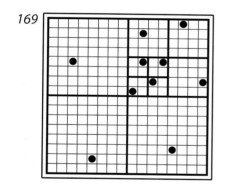

170
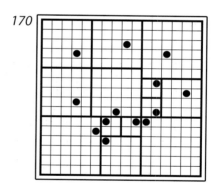

171
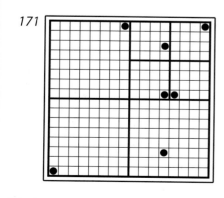

172
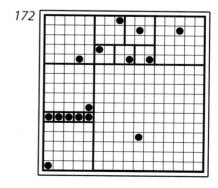

Solutions

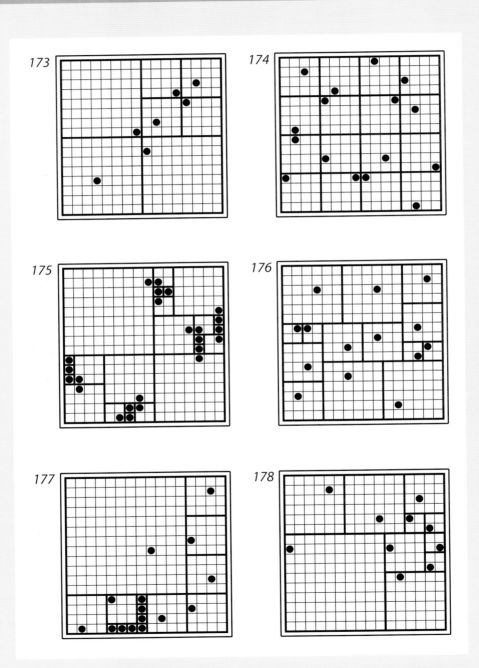

Solutions

Find the diamonds

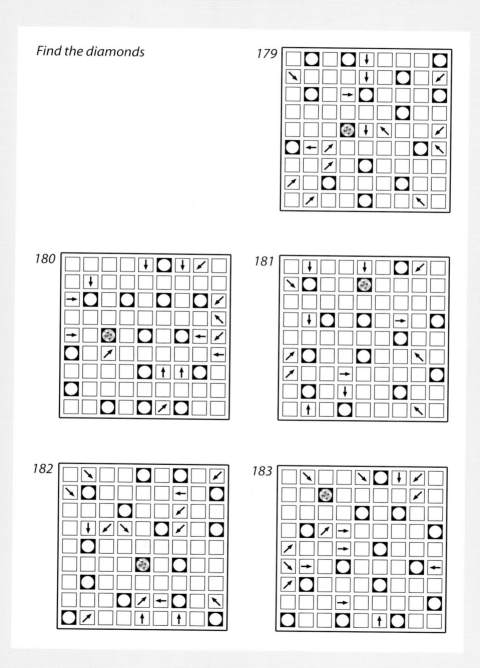

Solutions

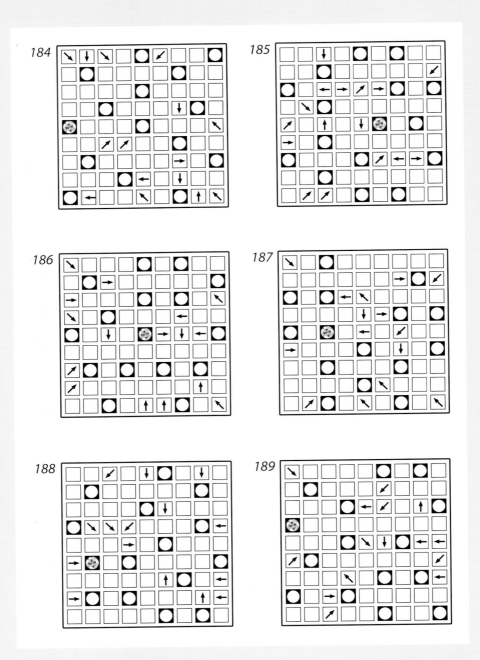

Solutions

Binary

190
```
1 0 1 0 0 1 0 0 1 0 1 1
0 1 0 1 1 0 0 1 1 0 0 1
1 1 0 1 0 0 1 1 0 1 0 0
1 0 1 0 0 1 1 0 1 0 1 0
0 0 1 0 1 1 0 0 1 1 0 1
1 1 0 1 0 0 1 1 0 0 1 0
0 0 1 0 1 1 0 1 0 1 0 1
0 1 0 1 1 0 1 0 1 0 1 0
1 0 1 1 0 0 1 0 0 1 1 0
0 1 0 0 1 1 0 1 0 1 0 1
1 0 1 1 0 1 0 0 1 0 1 0
0 1 0 0 1 0 1 1 0 1 0 1
```

191
```
0 1 1 0 0 1 1 0 1 0 0 1
1 1 0 0 1 0 1 0 1 0 0 1
1 0 0 1 1 0 0 1 0 1 1 0
0 1 1 0 0 1 1 0 0 1 1 0
0 0 1 1 0 1 1 0 1 0 0 1
1 1 0 1 1 0 0 1 0 1 0 0
0 0 1 0 1 1 0 1 0 1 1 0
1 0 1 1 0 0 1 0 1 0 0 1
0 1 0 0 1 0 0 1 1 0 1 1
1 0 0 1 0 1 1 0 0 1 1 0
1 0 1 0 1 0 0 1 0 1 0 1
0 1 0 1 0 1 0 1 1 0 1 0
```

192
```
0 0 1 1 0 0 1 0 1 1 0 1
1 0 0 1 0 1 0 1 1 0 1 0
0 1 1 0 1 0 0 1 0 1 0 1
0 1 0 0 1 0 1 0 1 0 1 1
1 0 1 1 0 1 0 0 1 1 0 0
0 0 1 0 1 0 1 1 0 0 1 1
0 1 0 1 0 1 0 1 0 1 0 1
1 0 1 0 0 1 1 0 1 0 1 0
1 1 0 0 1 0 0 1 0 1 1 0
0 0 1 1 0 1 1 0 1 0 0 1
1 1 0 0 1 1 0 1 0 0 1 0
1 1 0 1 1 0 1 0 0 1 0 0
```

193
```
0 1 0 0 1 0 1 1 0 1 0 1
1 0 0 1 1 0 1 0 1 0 1 0
0 1 1 0 0 1 0 1 0 1 0 1
1 0 0 1 1 0 1 0 0 1 0 1
1 0 1 1 0 1 0 0 1 0 1 0
0 1 1 0 0 1 0 1 1 0 0 1
1 1 0 0 1 0 1 0 0 1 1 0
1 0 0 1 1 0 0 1 1 0 1 0
0 0 1 0 0 1 1 0 1 1 0 1
1 1 0 1 0 1 0 1 0 0 1 0
0 1 1 0 1 0 1 0 0 1 0 1
0 0 1 1 0 1 0 1 1 0 1 0
```

194
```
1 0 1 1 0 1 1 0 0 1 0 0
1 1 0 0 1 1 0 1 0 1 0 0
0 1 0 1 0 0 1 0 1 0 1 1
0 0 1 1 0 1 0 1 1 0 1 0
1 1 0 0 1 0 1 1 0 1 0 0
1 0 1 1 0 1 0 0 1 0 0 1
0 1 1 0 1 0 1 0 1 0 1 0
1 0 0 1 1 0 0 1 0 1 0 1
0 0 1 0 0 1 1 0 1 0 1 1
0 1 0 1 0 1 0 1 0 1 1 0
1 0 1 0 1 0 1 0 0 1 0 1
0 1 0 0 1 0 0 1 1 0 1 1
```

Solutions

195
```
1 0 0 1 1 0 1 0 0 1 0 1 1 0
0 1 1 0 1 1 0 0 1 1 0 0 1 0
1 1 0 1 0 1 0 1 0 0 1 0 0 1
1 0 1 1 0 0 1 1 0 1 0 1 0 0
0 1 1 0 1 1 0 0 1 0 1 0 1 0
0 1 0 1 0 0 1 1 0 1 0 1 0 1
1 0 1 1 0 0 1 0 1 0 0 1 0 1
1 1 0 0 1 1 0 0 1 0 1 0 1 0
0 0 1 0 0 1 0 1 0 1 1 0 1 1
1 0 1 1 0 0 1 1 0 0 1 1 1 1
0 1 0 0 1 1 0 1 1 0 1 0 0 1
0 0 1 0 1 0 1 1 0 0 1 0 1 1
1 0 0 1 0 1 0 0 1 1 0 1 1 0
0 1 0 0 1 0 1 1 0 0 1 1 0 1
```

196
```
0 0 1 1 0 1 0 1 1 0 1 0 1 0
1 0 0 1 1 0 0 1 0 1 0 1 1 0
0 1 1 0 0 1 1 0 1 0 1 0 0 1
1 1 0 0 1 0 1 0 1 0 0 1 1 0
0 0 1 1 0 1 0 1 0 1 0 1 1 0
1 1 0 0 1 0 1 0 0 1 1 0 0 1
0 0 1 1 0 1 0 0 1 0 1 0 1 1
1 1 0 0 1 0 1 1 0 1 0 1 0 0
0 1 0 1 0 1 1 0 0 1 0 0 1 1
1 0 1 0 1 0 0 1 1 0 1 1 0 0
0 1 1 0 1 0 0 1 0 0 1 1 0 1
1 0 0 1 0 1 1 0 1 1 0 0 1 0
1 0 0 1 0 1 0 0 1 1 0 1 0 1
0 1 1 0 1 0 1 1 0 0 1 0 0 1
```

197
```
0 1 0 0 1 1 0 0 1 0 1 1 0 1
1 0 0 1 0 0 1 0 1 0 1 1 1 0
1 0 1 0 0 1 0 1 0 0 1 0 1 1
0 1 0 0 1 0 0 1 1 0 1 0 1 0
0 0 1 1 0 1 1 0 1 1 0 0 1 0
1 0 0 1 0 1 1 0 0 1 0 1 0 1
0 1 1 0 1 0 0 1 1 0 1 0 1 0
1 1 0 0 1 0 1 1 0 1 0 0 1 0
1 0 1 1 0 1 0 0 0 1 0 0 1 0
0 0 1 0 1 0 1 1 0 0 1 1 0 1
1 1 0 1 0 1 1 0 0 1 0 0 1 0
0 1 1 0 1 0 0 1 1 0 1 1 0 0
0 0 1 1 0 1 1 0 0 1 1 0 0 1
1 1 0 1 1 0 0 1 0 1 0 0 1 0
```

198
```
1 0 1 0 1 0 1 0 1 0 1 0 0 1
1 1 0 0 1 0 1 0 0 1 0 1 1 0
0 1 0 1 0 1 0 1 0 1 0 1 1 0
0 0 1 1 0 1 0 1 0 1 1 0 0 1
1 0 1 0 1 0 1 0 1 0 0 1 1 0
0 1 0 1 0 1 0 1 0 1 1 0 0 1
1 1 0 0 1 0 1 0 0 1 1 0 1 0
0 0 1 1 0 1 0 1 0 0 1 1 0 1
0 1 0 0 1 1 0 1 1 0 1 0 0 1
1 0 1 1 0 0 1 0 0 1 1 0 0 1
0 1 0 1 1 0 0 1 1 0 0 1 1 0
1 0 1 0 0 1 1 0 1 0 0 1 0 1
0 0 1 0 0 1 0 1 0 1 1 0 1 1
1 1 0 1 1 0 0 1 0 1 0 1 0 0
```

199
```
1 0 1 0 0 1 0 0 1 1 0 0 1 1 0 1
0 1 0 1 0 0 1 0 1 1 0 0 1 1 0
1 1 0 1 0 0 1 0 1 0 0 1 0 1 0
1 0 1 0 0 1 1 0 1 0 0 1 1 0 0 1
0 0 1 0 1 1 0 1 0 1 1 0 0 1 1 0
1 1 0 1 1 0 1 0 1 0 0 1 0 1 0
0 0 1 1 0 1 0 0 1 0 0 1 1 0 0 1
1 1 0 0 1 1 0 1 0 1 0 1 0 0 1 0
0 0 1 0 1 0 1 1 0 1 1 0 0 1 0 1
0 1 0 1 0 1 1 0 1 0 0 1 0 0 1 0
1 0 1 0 1 0 0 1 0 0 1 0 1 1 0 1
1 0 1 1 0 1 0 1 0 1 1 0 0 1 0 0
0 1 0 1 0 0 1 0 1 1 0 1 0 0 1 1
0 1 0 1 1 0 0 1 1 0 1 0 0 0 1 1
1 0 1 0 0 1 0 1 0 0 1 1 0 1 1 0
0 1 0 1 0 0 1 0 0 1 0 1 1 0 1 1
```

200
```
0 0 1 0 1 0 1 1 0 1 0 1 0 1 1 0
0 0 1 0 0 1 1 0 1 1 0 0 1 1 0 1
1 1 0 1 0 1 0 0 1 0 1 1 0 0 1 0
1 0 0 1 1 0 0 1 0 1 0 1 0 0 1 1
0 1 1 0 0 1 1 0 0 1 1 0 1 1 0 0
0 0 1 0 1 0 0 1 1 0 1 0 1 0 0 1
1 1 0 1 0 1 1 0 1 0 0 1 0 1 0 0
0 1 1 0 0 1 1 0 0 1 1 0 1 0 1 0
1 0 0 1 0 1 0 0 1 0 1 1 0 0 1 0
1 1 0 1 1 0 0 1 0 1 0 1 0 1 0 0
0 0 1 0 0 1 1 0 1 0 0 1 1 0 1 1
0 1 0 1 1 0 0 1 0 1 0 0 1 0 1 0
1 0 1 0 0 1 1 0 1 1 0 1 0 1 1 0
1 1 0 0 1 1 0 0 1 0 0 1 0 1 1 0
0 1 0 0 1 1 0 1 0 0 1 0 1 1 0 1
1 0 1 1 0 0 1 0 0 1 1 0 1 0 0 1
```

Solutions

201

202

203

204

205

206

Solutions

Page 9   *A hole*

Page 111   *1=0.999… or 1>0.999…*
There is a lot of discussion about this seemingly simple question. Surf the Internet and you'll find almost as many pro and con arguments.

PeterFrank t.v. was founded in 2000 by Peter De Schepper and Frank Coussement. The internationally registered trademark BrainSnack® stands for challenging, logical puzzles and mind games for kids, young adults, and adults. It also stands for high quality puzzles. Whether they are made by hand, like our visual puzzles, or generated by a computer, like Sudoku puzzles, all BrainSnacks® are tested by the target group they were made for before we put them on the market. In order to guarantee that our computer-generated puzzles can actually be solved by humans, we make programs that only use human logic algorithms.